# Playing in the FM Band

# STEVE POST

Foreword by Julius Lester
Illustrated by Ira Epstein

A PERSONAL ACCOUNT O

REE RADIO

# PLAYING IN
# THE FM BAND

THE VIKING PRESS   NEW YORK

*This book is dedicated with love and gratitude*
*to my father, Nathan Post*

First published in 1974 by The Viking Press, Inc.
625 Madison Avenue, New York, N.Y. 10022
Published simultaneously in Canada by
The Macmillan Company of Canada Limited
SBN 670–55927–x
Library of Congress catalog card number: 73–8688
Printed in U.S.A. by H. Wolff Book Mfg. Co.

*As an act of charity, certain names have been changed*
*and situations slightly altered.*

ACKNOWLEDGMENTS

*Broadcasting:* Reprinted with permission from the
August 23, 1971, issue of *Broadcasting* magazine.

*Long Island Press:* From articles that appeared
in the December 3, 4, and 16, 1970, issues.

*The New York Times:* From March 6, 1972, editorial,
"The Goodman Jailing," and Letter to the Editor by
Frank Millspaugh of March 6, 1972.
© 1972 by The New York Times Company.

*Newsday:* Excerpts from articles that appeared on
December 8, 1970, and January 4, 1971.
Copyright © 1970 (1971) Newsday Inc.

*Pantheon Books:* Excerpts by
Christopher Koch and Lewis Hill from *The Exacting Ear,*
edited by Eleanor McKinney.
Copyright © 1966 by Pacifica Foundation.
Reprinted by permission of Pantheon Books, a Division
of Random House, Inc.

*The Village Voice:* From "Son of Play List: The Decline
and Fall of Commercial Free-Form Radio" by Steve Post.
Copyrighted by The Village Voice, Inc., 1971.
Reprinted by permission of *The Village Voice.*

# Thanks

I have often suspected that only those who expect to find their names in this section bother reading it. Yet I must admit that naked curiosity compels me to continue reading through such lists of unfamiliar names, perhaps in the vain hope of finding myself among those acknowledged, though it hasn't happened yet.

The compulsive acknowledgments reader must, upon occasion at least, wonder just how it is possible that so many have contributed so much to so little. But the other end of the pen, like the other end of the microphone, provides fresh perspective, and it is now clear to me that any work, even one confined to so narrow a topic as this, is the result of a lifetime of associations. The act of spelling out their names with a blob of ink on a piece of paper seems ludicrously inadequate. I am certain that I will leave out many names that should be included, and I suspect that some who have been included would prefer to be left out.

Sprinkled conspicuously through this book are a couple of letters written to me by Ira Epstein. At the outset, I would like to disassociate myself from both Ira and his letters. They would not even appear here were it not for the persistent, annoying telephone calls from Ira to me and my editors at all hours of the day and night. Finally we gave in. My personal advice would be for you to treat them as you would a TV or radio commercial; when one appears, simply run out and grab a snack, or relieve yourself of an earlier one. I fear, however, that if you are curious enough to plow through these acknowledgments, the same curiosity will lead you at least to skim through Ira's works. So I will tell you how they came to be.

One day during the fall of some year in the late nineteen-sixties, I found in my station mailbox a neatly typewritten letter illustrated with pen-and-ink drawings. The contents of the letter were in bad taste and personally offensive. I read it on the air. Encouraged by an indifferent response from the audience, Ira expanded that letter into a series. Over the next several years, right up until the end of my late-night program, *The Outside*, I received a steady, if irregular flow of such junk.

The letters came for almost a year before I had occasion to meet my correspondent. Then, one night in November, on an anniversary for *The Outside*, there appeared in the control room a tuxedoed young man who resembled a penguin at commencement exercises. He introduced himself as Ira Epstein. Not wishing to be impolite, I extended my hand and informed him that I loathe having people in the studio when I'm on the air. But Ira was already well into pouring out the story of his life. I tried spitting on his cummerbund, but he didn't seem to notice. Before long I found out that this pathetic young man intended to go to medical school (God help us) and was now working his way through his final year of college by playing an assortment of musical instruments at catered affairs. Earlier that very day he'd played at a wake and managed to slip out with the deceased for later use in his medical training. (He is, by the way, available for your next affair and will, for a fee, play a medley of every song ever written on the instrument of your choice.) [For Ira's vision of this encounter, see pages 57–61.]

I would like to thank Ira from the bottom of my heart for his vast contribution to this work, to my entire life, and to the plight of the sick and infirm the world over.

The following people provided encouragement and/or inspiration in the treacherous crossing from the spoken to the written word: Paul Krassner, Jeff Greenfield, Ian Whitcomb, Julius Lester, and John Leonard, from *Nightsounds* to *This Pen for Hire*; my editors at The Viking Press, Linda Yablonsky and Barbara Burn, who naïvely took for granted my ability to produce this work, even as months passed with no evidence to back that assumption; special thanks (or blame) to Linda Yablonsky, who initiated the idea for this book, which otherwise would have remained filed along with my thousands of other unfulfilled fantasies.

It would take several volumes to adequately thank the incredible assortment of Pacifica people who are, in effect, the author. More than anything else this book is their story, and I suspect that most of them could have done a far better job of telling it than I have. In a normal life span one would have to be extraordinarily fortunate even to make the acquaintance of such a bizarre, perceptive, and talented group. I can't begin to name them all, but here are a few who have kept listener-sponsored radio alive and vital, often against overwhelming odds, and who along the way have profoundly influenced my life: Bob Fass, Larry Josephson, Frank Millspaugh, Dale Minor, Elizabeth Minor, Bill Butler, Kathy Dobkin, Milton Hoffman, Robbie Barrish, David Rapkin, Neil Conan, Nanette Rainone, Peter Zanger, Marv Segelman, Florrie Segelman, Bridget Potter, Paz Cohen, Margot Adler, Frank Coffee, Ed Goodman, Elsa Knight Thompson, Larry Lee, Liza Cowan, Chris Koch, Marc Spector, Charles Pitts, Andy Getz, Paul Gorman, Marnie Mueller, Don Gardner, Bill Schechner, Bob Kuttner, Robert Goodman, Dr. Caroline Goodman.

During the eight years of *The Outside*, fewer than two dozen guests appeared on the program. Most appeared only once, but a few, foolish enough to return, became the family. They are an extraordinarily gifted, weird bunch. First, the incredible Marshall Efron: we had a perfect fifty-fifty partnership—he provided the talent, I provided the temperament; genius is not too strong a word to describe that talent. Paul Krassner, forty-plus-year-old alienated youth, who made me feel secure in my indiscriminate laughter. Marilyn Sokol and her friends Smootchy Farcas and Sara Goy. The uplifting Brother Theodore, who is cursed with a unique talent. Ian Whitcomb, musician, wit, world traveler, author, and friend. And Barton Heyman.

A special thanks, of course, to my six listeners, who, had they not been total social outcasts, most surely would have found something better to do on Saturday and Sunday nights.

The players in the C. W. Post drama: Jackie Reubens, Kent Mosten, Mark Indig, Jane Bletcher, Mel Rosen, Big Ray Mineo, Bobby "G," Dr. and Mrs. James Dwyer, Dr. Herb Kosten, Patti Baltimore, William Kunstler and the staff of the Center for Constitutional Rights, Alex Bennett, Ronnie Bennett, and the thousands of students, faculty, and media people who supported us against impossibly powerful odds.

My stepmother, Madeleine Post, who, among other things, encouraged me through the first awful pages of the first draft. And my brother, Gerald Post ("Dee Dee"), who, thank God, absorbed some of the genetic insanity.

And finally those few special friends, whose contributions are too vast and varied to be pinned down: Mady Burton; Frank Millspaugh; Robin Fitelson; Paul Fischer; Laura Rosenberg, who looked at and lived with me and the book through most of it, undoubtedly the most hideous of all tasks; and James Irsay.

# Foreword

Although hundreds of books and articles have been written about television, its uses, misuses, and overwhelming impact on our lives, radio has been consistently ignored by social analysts and commentators. This is not surprising, because a radio does not sit in the center of our homes staring back at us like a squat Cyclops, transfixing and mesmerizing us in the glow of its eye. Radio does not occupy the center of family life, with its programs the major topics of conversations, its presence more immediate than those with whom we live.

Yet radio is as much a part of our lives as television. Unobtrusive, quiet, almost invisible, it is there, and wherever we go, more often than not, a radio goes with us—along expressways and streets, to beaches, parks, supermarkets, and shopping centers. Most homes have several radios, at least one of which is next to the bed, where it awakens us each morning with news and music. Radio is so integral a part of us now that we do not consciously notice its presence; it is a member of the family, a companion, and the voices issuing through its speakers are those not of strangers but of friends.

Of all the electronic media, radio is the most intimate. The disembodied human voice has a power and mystery wholly unavailable to television's images. Those of us who grew up in those ancient days before television can remember in the very fibers of our nerve endings the sound of the creaking door and the tremor in the announcer's voice as he said, "Good evening, friends. This is Raymond, your host, welcoming you to the Inner Sanctum. . . ." Each of our minds created its own vivid pictures of the house, castle, room to which the door was attached by rusty hinges. Unlike television and the movies, where the drama takes place outside of us, radio made us an equal participant, and when Sergeant Preston of the Yukon shouted, "On! On! On, you huskies!" we cracked the whip with him and hunched our shoulders against the cold Arctic winds blowing snow into our faces as we clung to the runners of the dogsled.

In the days before television, radio was merely an electronic extension of the times when entertainment was provided in the home and community by story-telling and singing. Though radio took that away, it did not turn us into passive victims of electronic massages. Radio demanded our participation, and even after we knew that the sound of the wind whistling around the house on *I Love a Mystery* was nothing more than a giant fan, it didn't matter. We were participants in an act of imagination.

With the advent of television, radio declined for a period as its programs were adapted for the video screen. In its search for a new definition and function, radio eventually settled upon news and music, and there, with few exceptions, it has remained today. Though it is seldom the medium for the imagination that it had been, it has retained its intimacy in our lives. Sadly, however, it has ceased to be an adventure. Primarily this is because the majority of radio stations are businesses like any other business, and their primary concern is not the product but whether the accountant's books are filled with solid, respectable black ink or frightening red.

Steve Post's *Playing in the FM Band* is the story of a radio station whose programing is a constant adventure, and whose day-to-day fight for survival is often one as well. WBAI-FM (99.5) in New York City is the flagship station for the Pacifica Foundation, which also operates stations in Berkeley and Los Angeles, California; Houston, Texas; and Washington, D.C.—if the Federal Communications Commission approves Pacifica's license application. These stations are unique because their sole support comes from their listeners, and this is both the weakness and strength of the Pacifica adventure. Because its programing cannot be affected by threats from commercial sponsors to withdraw accounts if a certain program or air personality is not removed, "those of us who broadcast over listener-sponsored, Pacifica-owned stations," Post writes, "say what we want to say, when and how we want to say it, as long as it is not in direct violation of Federal Communications Commission (FCC) regulations. And even then we have been known to attempt to broaden the interpretation of those regulations." This kind of freedom is available, or so it seems, only through being wholly listener-sponsored, and the price of this freedom is poverty, because more people listen than sponsor.

Yet the concept of listener-sponsored radio is audacious and inspiring, because, in a democracy, the Pacifica stations are truly of, by, and for the people. Steve points out that "it was the intention of Pacifica's founders to develop a radio station that spoke to the minority. They believed that in a society which supposedly guaranteed the right of all its citizens to freedom of speech, no matter what their views, there should be free and open access to the electronic media as well." Under commercial sponsorship, radio is not able to fulfill this function. Ford Motor Company will not sponsor radio programs allowing Black Panthers, the Women's Liberation movement, or homosexuals to express their views. The Bell Telephone Company will not pay for a two-hour concert of avant-garde music, a reading of Greek poetry, or a discussion by film directors. Commercial radio, as Steve says, must sell products and to do so "must cater to the lowest common denominator of intelligence."

Thus the Pacifica stations and WBAI in particular, which this book is about, are unique in America and perhaps the world. That is not something of which any of us should be proud. It is a sad commentary that in the Greater New York listening area, covering the five boroughs of New York City, New Jersey, parts of Connecticut and upstate New York, there is only one WBAI out of more than one hundred stations broadcasting to a population upwards of ten million. But to be a WBAI means taking risks, the first of which is that the listeners will not send in enough money each and every year to pay staff salaries, buy tape and editing machines, keep the transmitter in working order, pay the bills for electricity, telephones, the lines necessary for remote broadcasts, and on and on and on. That is only the initial risk, however, because a station like WBAI, to be alive and vital, must live in a state of constant jeopardy, always doing that which other stations cannot or will not do. Inevitably that means giving air time to people and organizations with unpopular views, be they from the right or left of the political spectrum. The Ku Klux Klan and Black Panther Party have an equal right to be heard through the medium of radio, and WBAI has made its air available to both. Flag-wavers and flag-burners should be equal in the eyes of the media; unfortunately, they are not, particularly during a time when the administration in Washington is making a determined effort to transform the media into a wing of government. The First Amendment guarantees free speech, but free speech is not a concrete entity existing in and of itself. Free speech exists only if it is practiced; a free press is only as free as it will force itself to be, and in America that is not very free.

This book is Steve Post's personal account of an institution that is fiercely dedicated to free speech. Of the hundreds who could have written about WBAI, Steve is perhaps the best to have done so, because for seven years WBAI was his life, and it is important to understand that BAI is as much a way of life as it is a radio station. Steve was a high-school graduate who, like many others, had little idea of what to do with his life. After unsuccessful attempts to become a part of the workaday world, he stumbled onto WBAI and was made the station's bookkeeper. This was in spite of the fact that Steve has been known to make a mistake while counting on his fingers. But, as Steve recounts, being the bookkeeper for BAI didn't have as much to do with math as it did with holding off creditors. However, the mere fact that he was given such a job is instructive about the institution itself, which does not function by job interviews, personnel committees, or, obviously, job criteria. Rather, there seems to be an invisible antenna which scans the souls of those who come to the station, which looks for people who have not squelched and buried whatever it is that makes them unique.

Steve was unique and, like most people, did not know it. Unlike anybody at the station before or since, Steve's way of approaching the world was humor. "I had learned early in life to use humor as both a defense and a tool, to understand why I could not adjust, as could so many of my contemporaries, to the demands imposed on me by a set of values that seemed to have no distinct relationship to my own feelings and needs." Yet, when he mercifully moved from the bookkeeper's desk to the chair behind the microphone, he did not recognize that this was his uniqueness. Then again, perhaps there are few things in the world more terrifying than having the responsibility of a live radio show, particularly at WBAI.

Steve goes into great detail here about the evolution of what is now called "free-form" radio, i.e., live radio where the program producer is free to do whatever he wants—play records, talk, take telephone calls on the air, eat his dinner, belch, etc. As Steve says, and quite accurately so, free-form radio is an art form. The airwaves are the empty canvas, the producer is the artist, and sound is the paint. WBAI's Bob Fass was, more than any other person, the creator of this form, and Steve writes movingly of sitting in the studio, watching Bob do his midnight-to-dawn show. "Fass would arrive only moments before the broadcast, loaded down with shopping bags, briefcases, and cartons filled with records and tapes. . . . *Radio Unnameable* proceeded from moment to moment, with Fass free-associating one record, tape, or thought to another, generally deciding upon the next record or tape only as the previous one ran out. He would open the microphone and talk about something he had seen, heard, or done, then follow it with a string of recorded material relating either directly or abstractly to that theme. Then, just as spontaneously, he would juxtapose other thoughts and begin a new theme, read a piece of poetry, or produce an unprepared collage. Listening at home, one could never know what to expect next. And that, I soon learned, is an important part of the magic of free-form radio. . . . *Radio Unnameable* had a very real and direct relationship to its audience, something lacking in the rest of the media, and even in most of WBAI's programing."

Fass understood the incredible intimacy of radio, and it was something Steve was going to have to learn, slowly and painfully.

Free-form radio requires that the producer (the air personality) be open and vulnerable to the human experience of which he partakes, and that his show speak from his own condition as a vulnerable human to that same condition in his listeners, because that is the basis of existence which all of us share. During his first show, Steve writes, "I painfully experienced the technical and emotional barriers inherent in this new and different use of the medium. They are barriers unseen and unfelt by the audience. Barriers that, should they be detected, lead to the failure of the effort, the breaking of the spell." He continues, eloquently, to go to the heart of not only the problems he faced but the problem of radio itself: "How is it possible to communicate experience and feeling isolated from those to be communicated with, shut off in a tiny room absent of living matter, staring at clocks, meters, and machinery, pushing buttons, pulling levers, surrounded by glass and metal? Who are those faceless beings we call 'listeners'? What are they doing? Lovers in the act of love-making? The desperate, who in their despair have found a final, comforting, undemanding companion to speak to them? Those who want simply to be 'entertained,' enlightened, or amused? And where, while dealing with such realities as finding the next record or tape to play, is there time even to consider such things? And what is it, finally, that gives me the right to be on this end of the microphone?"

It was only after a year and a half on the air that Steve began to answer these questions in his own unique way, but when he did, the result was one of the most special radio shows in the country, and for those of us who listened every Saturday and Sunday night, Steve's present absence from that time slot is an irreplaceable loss. At a radio station which leaned so far to the left that it was almost horizontal, Steve Post became the necessary antidote, not because he was a conservative, but because he held up a different mirror to our faces, one that did not reflect the image we were accustomed to seeing. He dared to satirize blacks, women, peace marchers, and anyone else he felt was a little too self-righteous about a particular grievance. He dared to satirize the seriousness of telephone call-in shows by being rude to callers and never hesitating to put down the pompous intellectuals, ridicule the too serious, and hang up eagerly on the boring.

Beyond that, however, his show was important because it was the only place on radio one could hear live, spontaneous comedy. Steve was not a monologist like Jean Shepherd, re-creating his childhood, but a poor man's Woody Allen, life's whipping boy, acutely aware of his limitations, and by verbally playing with them, he made us who listened to him more accepting of our own limitations. "We are all ridiculous human beings," his shows seemed to say, "struggling with this gargantuan force called life ineptly and ineffectively and making asses of ourselves in the process. Let's stop being so serious about it." Because he knew his own weaknesses, his instinct was for ours, but the difference between Steve and us was that he accepted his. The night he told the story of walking into a pizzeria with a Procaccino-for-Mayor button on and having the Italian counterman laugh at him uproariously, all the while calling all the other Italians in the place to look, Steve was not only telling a story of which he was the butt, but cutting away at our own self-pretentious attitudes, putting us in touch with our own humanity.

It was healthy to have in our midst someone for whom nothing was so serious that it couldn't be laughed at. Listening to Steve's show was very good for me, a serious, intense person, afraid to look at my frailties. At first I was uncomfortable listening to him, hating him, but unable to turn to another station on Saturday and Sunday nights. Eventually I realized that I was being not ridiculed but played with. Steve tells the story here of how listening to Larry Josephson's morning show influenced and shaped what he did on the air, and for me, Steve taught me more about my own uniqueness and how to express that on the air than any other person. That is another story, but I mention it here because Steve taught many of us something about being human. His show was very serious, but only in retrospect do I recognize that. It was easy to laugh at Steve telling us of his undying crush on and love for the singer Judy Collins, and only later, when the laughter had died, did you realize that he had been talking, not about himself but about you and the secret crushes we carry for public personalities. (I have one on Jackie Onassis.) Comedy has within it the ability to affect lives profoundly, and, consciously or not, Steve knows this and acts upon it.

This book, then, is the story not only of one particular radio station, but also of a particular individual's experience as a part of it. As such, it is important and vital, because it also tells the story of what radio can be when unfettered by accounting books, when it is almost monomaniacal about free speech. Freedom, like love, is an experience that must be constantly lived to exist. We are only as free as we act. WBAI is one of the few examples of freedom in action in the electronic media.

I'm glad that Steve went through the agony of writing this book to share with us his and WBAI's stories. Since I have been a part of that station since 1968—ironically enough by stumbling into the station one night and being asked by Bob Fass if I'd like to sit in for Steve, who was away—it is my story too, as a staff member and as a listener. It is also your story, those of you who have listened to the station and supported it. It is also the story of those of you who may be stuck with top-forty and all-news radio stations, because it tells what can be done, what should be done. Basically, this is a book about your freedom and your lives.

JULIUS LESTER

# Contents

# Between Fond Recollection
# and Bitter Memoir:
# Introductory Fragments

In January 1971 I was lying in a hospital bed recovering from the second in a series of operations for the removal of a growth at the base of my spine. I had entered the hospital at the very beginning of the month for what, I was told by the doctor, would be a fifteen-minute operation and two or three days of recuperation. By mid-January I'd spent nearly four hours on the operating table and more than two weeks lying on my right side recuperating. Things were not looking up.

Two and a half months earlier I'd been fired from my job as Director of Radio at C. W. Post College. The week before that I had been voluntarily laid off from my weekend radio program on WBAI to help ease the station's financial burden. After all, I had reasoned, I was in a financially secure position at the college. I've always had a magnificent sense of timing. Now here I was, unemployed, hospital bills accumulating like whale blubber. Our legal case against the college was still pending on appeal, but we had already lost several rounds both in the courts and on campus, and it was as obvious as Richard Nixon's five o'clock shadow that we would lose this final round. It never rains but it floods.

Lying in that hospital bed, semilucid from painkillers administered through existing and freshly created orifices, I made several barely coherent on-the-air telephone calls to WBAI, describing in unnecessarily graphic (or so I've been told) detail my medical problems, with particular emphasis on my battle with three burly nurses over their apparent desire to subject me to an enema—a procedure I had managed to resist throughout my previous twenty-six years of life. Somehow death through implosion seemed preferable.

On January 21 I received a three-page letter at the hospital from an Assistant Editor at The Viking Press, which began: "I chanced to hear your phone conversation with Paul Gorman on BAI last Saturday afternoon, and I want to offer you my very deepest empathies and an equally sincere proposition. . . ." Already it sounded good.

The writer then chronicled, in exquisite detail, her own adventures with a similar ailment. That letter itself is worthy of publication, both as a pilot script for a TV situation comedy and as a paper on the ineptitudes of the medical profession.

Finally, the proposition. Hardly what I had expected (or hoped for): "But even if things don't turn out completely rosy, I think you would be an excellent person to write a book on the subject of 'underground' or 'alternate' radio, with emphasis on the college and listener-sponsored type stations and programing. . . . I figure you could discuss your experiences, say with WBAI and with other Pacifica stations, and the whole C. W. Post thing. . . . I put this to you without knowing whether or not you can write well at all—this is just something that occurred to me on Saturday as I listened to you."

Very odd. For years I'd had a fantasy—strictly a fantasy—of doing a book about my life at WBAI. My six years with the station had been not only an experience of growth, but one of creation, of birth. But I had set down not one word. Nor did I intend to. It was as distant a fantasy and impossible a task as proving the world to be flat. Walter Mitty would not even have dared to daydream about this one. At the age of eight I developed writer's block midway through a grade-school composition on "What I Did During Summer Vacation." Since then I had required a ghost writer to fill in my annual *Life* magazine subscription renewal form. "A book? Me? Are you kidding?"

But others I knew who had, might have, were, would, or should be writing books suggested that, capable or not (the latter forming the consensus), I ought not dismiss the offer lightly. Publishers, after all, present potential authors with an "advance against royalties," and rarely do they institute legal action to recover moneys from those writers who don't come through. Such are the hazards of the publishing business, and besides, a tax write-off, while not the finest of literary creations, is often more profitable, and even more interesting, than the story the advance intended to tell.

And so, once out of the hospital, while continuing to recuperate on my right side, I began working up an outline for this book. I don't know how many of you have ever tried typing while lying on your right side, but that feat alone seems worthy of a Pulitzer Prize. Seventeen thousand packages of Ko-Rec-Type later, I completed that outline, which sounded something like the conclusion to my grade-school composition, though it might have been a little late to satisfy Miss McGlynn, my blue-haired teacher. Still it seemed appropriate, since I've tended to view life at WBAI as somewhere between summer camp and primal-scream therapy.

Before too long, and to my utter disbelief, I had in my possession a book contract from The Viking Press, and a piece of that advance, promptly deposited in a savings bank. A Xerox of the check hangs on my bulletin board, an almost-living monument to the gullibility of the publishing industry. They were more easily conned than dear Miss McGlynn. To paraphrase Bob Dylan, I planned to sit by my savings bank and watch the interest accrue.

But two nagging problems persisted. One was the deeply ingrained belief that I would be either "left back" or, worse yet, taken from my home in the middle of the night by a Viking Press truant officer and placed in a bare room with only a typewriter and a stack of blank yellow paper, capable of reproduction at a rate roughly equivalent to that of laboratory mice. The other, and in the end the more compelling, was a seemingly inescapable need to set it all down so that, if nothing else, I would have something in print to pierce the serene senility of later years, to see how I occupied the third decade of my life. Both are admittedly selfish motives.

I felt at the time that WBAI had done and could do no wrong, at least as it affected my own life. Six years had not soured my view of the station. I knew others had left in bitterness, or worse, and I could speculate, at a distance, upon a similar fate for myself. But for now I began to tackle the task with the blind belief of a Jesus Freak on Sunday morning. Had the project been accomplished in the nine-month original deadline it probably would have read like an authorized history of Scientology written by a "clear." But as it developed, the next two years at WBAI, and in my life (barely distinguishable categories), proved to be critical ones. The manuscript was started, laid to rest, and started again. The process repeated itself. Each time I returned to it, it seemed outdated and inadequate, like yesterday's news, yesterday's insights, yesterday's dinner. And so at each of these points I set down some "Notes for an Introduction," an explanation, an excuse, an apology for what I had written. I hoped that in the end they would tie together. But they didn't, and to manufacture a thread on which to string them would be to deny the evolution of my own thinking in regard to the station and my own role within it. They were written in assorted emotional states. For what it is worth, I present them here as the fragments they are, with the approximate dates of their writing.

AUTUMN 1971

*To define listener-sponsored radio is all but impossible, since there is no particular goal, other than an articulated addiction to the First Amendment. Because WBAI is, more than anything else, a group of people who cling fiercely to their uniqueness, who work at the same time with and against each other, who love and hate, and build and destroy, all in the name of an elusive goal,* it cannot be said in a few words, or in an entire book, what WBAI is. If such a momentarily satisfying definition could be arrived at, it would only serve as a historical reference point by the time the galleys arrived. Which is not to say there are no constants at WBAI—there are: I myself seem to be one of them. But the constants are only superficially constant, forced to grow and change by those coming in the door with new ideas, energies, and enthusiasms. Perhaps the only real constant is the tension between tradition and innovation—a balance which must be struck, I believe, if the station is to survive.

And so I will introduce this book ass-backward, by telling you what it is not, and perhaps by the process of elimination we will figure out just what it is. Some will say even this is perfectly consistent with Pacifica tradition.

This book is not a definitive history of Pacifica Radio, nor even of WBAI. It is not a political or philosophical analysis of the institution. It is not a guide to programing or a manual on how to run a listener-sponsored radio station. It is not a dissertation on WBAI's role and influence in the media, nor of the media's role and influence on our lives and times. It is not my autobiography. It is not an anchovy or a vacuum cleaner.

I have spent more than half of my adult life at WBAI. Notice I don't say "working" at WBAI. It is not nearly so simple as that. For most of what I believe, feel, and do, now and for most of the past decade, the credit or blame can be placed directly on WBAI, or more accurately, on the people and ideas I have encountered there. I am a "Pacifica Orphan"—one of many who arrived at the front door with nothing in particular ready to offer—no innovative ideas, no apparent talent, no well-defined politics or ideology. Lost, a bit frightened by what we knew of the outside world, but eager, we were taken in, or slipped in, or pushed our way in. Some of us held on and found our place, or created one, and have become momentarily blind to the deficiencies of the institution. Others didn't make it, and left, disgruntled, angry, and have attacked the station, because reality failed to match expectation. The point is, don't expect to find "the truth" about WBAI in these pages. Nothing of the sort exists here. It is only a wholly subjective view of one person's incomplete experience.

---

It is amazing what changes a person goes through in the course of a project such as this, and no less amazing are the changes an institution such as WBAI experiences. Not only is it directed by the changes of those who remain with it, but it must also incorporate the passions, politics, and programs of the newly arrived, who often arrive with more passion and politics than programs.

My feeling now toward both my work and the station are ambivalent. If Lewis Hill, Pacifica's founder, were to listen to the station today he might roll over, if not spit up, in his grave. But then, it was Pacifica that put him there in the first place. Hill committed suicide in 1957. One institutional legend has it that he did so over a musical programing dispute. Another, probably more accurate explanation (though I'm convinced it is still not the whole story), claims the act was prompted by his disappointment and bitterness with the early political and financial struggles within Pacifica. And his was not an isolated example of self-destruction at least superficially traceable to Pacifica. Lew simply set the standard.

Today's WBAI is far different from the one I began writing about more than a year ago, as that one was from the one a year before it, and so on. And with the entrance of each new era there is no small opposition to it. This time the political and programing changes were preceded by what seemed like innocent and necessary physical changes. WBAI has gone from a rented floor in a brownstone with one tiny studio and a low-powered transmitter around the corner in a hotel, to owning (along with the bank) a large, reconverted church filled with shiny new equipment, several studios and production areas, and a fifty-thousand-watt (maximum legally permissible) stereo transmitter and antenna, "high atop the Empire State Building," as they say. Our signal is now as powerful and our sound as clear as any FM station in New York.

But what must really be evaluated, of course, is the content of that strong, clear sound. As society in recent years has seen the dissolution of the Movement into fragmented, militant special-interest groups—women, blacks, homosexuals, ethnic minorities, etc.—each with its own set of demands for participation and power, so WBAI has seen, felt, and been moved by this polarization.

Today the station is less the "family" I have referred to than it was in years past. Partly this is the inevitable, though saddening by-product of growth. With a staff of nearly fifty and volunteers running in the hundreds, representation at the station and on the air has gone from artistic to ideological. Which is not to say there have never before been ideological differences among staff members—on the contrary—but those differences remained secondary to the interests of creating exciting, provocative, and listenable radio. And though all varieties of ideological axes have been ground against WBAI's microphonic whetstone, it was not until recently that these grinders were placed on WBAI's payroll.

WBAI has, to an extent, become ghettoized. Personnel and program decisions seem as often to be made on the basis of politics as potential programing talent. And those who do remain with a desire to create purely beautiful radio programs must themselves organize into political alliances to protect their interests.

It is difficult to place an absolute value judgment on this kind of politicized programing and decision-making. Certainly it fills a need: those voices systematically unrepresented in the rest of the media roam freely over the airwaves of WBAI. Hours of air time are given over to ideologues of the feminist, black, and gay movements, or at least those segments of those movements represented by the programing staff and volunteers. All issues are politicized: a woman is fired from the station because her program is judged artistically inadequate, and pickets from a group calling themselves "Radical Dykes" parade in front of the station claiming she was dismissed because of her articulated lesbianism, though she herself stated her belief that the decision had to no degree been influenced by that factor. This kind of internal and external political pressure has become enormously influential in the WBAI decision-making process, often, I believe, to the detriment of programing.

----

MARCH 1973

The other day, during one of my increasingly infrequent off-the-air moments around the station, Ron Magee, a staff member of recent vintage, came into a studio where Paul Fischer, Marc Spector, and I had just finished a fund-raising appeal. "One thing I don't understand," he said, "is why you guys talk about this place like it's your mother."

Steve Post
RHINEBECK, NEW YORK
MAY 1973

# Playing in the FM Band

# 1

# *All Power to the Transmitter*

It was a Saturday night early in September 1967, late into another forgettable edition of *The Outside,* my regular late-night weekend radio program on WBAI-FM, New York. I was on the air, talking about something which has long since faded into electronic obscurity, when suddenly the voice coming through my headphones was no longer my own but that of another late-night radio personality on another radio station.

It was a startling but not unfamiliar occurrence—transmitter failure. WBAI was off the air. I'd been warned to be prepared for it, since we'd been having frequent and somewhat suspicious interruptions in our broadcasting for several weeks. The engineering staff had been conducting "field strength tests" to determine whether or not other stations were running more than the legally permissible amount of power, thus interfering with our signal.

Something in the past few weeks had been tripping the "fail-safe" mechanism in our transmitter, and suspicions of sabotage had grown during the last week when the transmitter failed several nights in a row just prior to news time. One member of the engineering staff, a new, part-time employee, claimed to have received several phone calls threatening the station. Paranoia was running high.

The station, as usual, was all but deserted that Saturday

night,* so I turned to the section on emergency procedures written by the chief engineer in the control-room instruction book. Not unexpectedly, I was unable to decipher them. (Radio station chief engineers are a universally odd lot. Like surgeons, they possess knowledge and skill which makes them nearly divine. And they know it. That is their power, and they are reluctant to share that power, even in a written manual designed for emergencies.) I phoned Tom Whitmore, our chief engineer, but there was no answer. I tried the numbers of a couple of others on the engineering staff. They too were out.

I left the lifeless control room to make one final check of the engineering office before I dared call Frank Millspaugh, the station manager, who I knew would not appreciate a phone call to his home at three in the morning. As I rushed down the hall leading from master control I bumped into the young engineer who had reported receiving the threatening calls earlier in the week.

"Thank God you're here," I said, more than a bit relieved. "We're off the air again."

"I know," he replied, slightly out of breath. "I was listening and I came running right down."

I asked if he'd been to the transmitter, in the Empire State Building, only a few blocks away from our East Thirty-ninth Street studios. He said he hadn't, so I asked him to run over there and call me as soon as he arrived.

He left, and I returned to master control to wait for his call, comforted that, at least, things were in competent engineering hands.

A half hour later I'd still heard nothing from the young engineer. I phoned the transmitter room, but the line was busy. A few minutes later I called again. This time the operator cut in and announced that the number was temporarily out of service. Being a recording, she could offer no further details, amplifying my vivid fantasies of a vast conspiracy against the station and, no doubt, me.

Finally, an hour later the engineer called. Relieved, though an-

---

* Recollections of the event differ. At least two engineers claim to have been present when the transmitter failed, but I distinctly remember searching the premises and finding no one about.

noyed, I asked why he had taken so long to phone.

"They've ripped the telephone from the wall," he answered, and in a shaky voice went on to describe in some detail a scene of vandalism at the transmitter. The door to the room, made partly of glass, had been smashed, he said, the transmitter itself broken into and a number of its vital parts removed or destroyed. Complex wiring had also been torn apart.

It seemed incredible. Though WBAI had received hundreds, perhaps thousands, of threats against its facilities and personnel during its seven years as New York's unorthodox listener-sponsored radio station, never before had it been physically attacked.

I phoned the station manager then, as well as the police, and remained in the control room to answer the hundreds of phone calls coming in from listeners—and, before long, from the press— meanwhile frightened ("scared shitless" would be more accurate) that the saboteurs might strike the studios next. But there were no further incidents—at least not that night.

When the damage was surveyed the next day, it became evident that this had been no random act of vandalism. While the destruction of the transmitter had not been extensive, it was precise. Essential elements had been carefully removed or tampered with— elements of no great expense, but those it would take days, or weeks, to replace. Intricate wiring would have to be redone, requiring days of nonstop work by skilled engineers. Whoever had visited the Empire State Building that night was no mischievous tourist angered by the nonexistent view of New York's polluted skyline. Our visitors were obviously familiar with radio transmitters and seemed especially well acquainted with WBAI's. A closer look at the door to the room showed that, despite the broken glass, the lock had been opened with a key.

It was a major news story the following day, with newspapers, radio, and television running lengthy accounts of the story. It is not every day that a radio or TV station is violently forced off the air, and when the station is WBAI, call letters associated with controversy—more often than not political—it is a major news story. Some accounts hinted that "right-wing extremists" might be to blame, a theory not completely discounted by those of us at the station.

But within the station there were growing suspicions of an "inside job," and police investigations soon centered around the young engineer, who, curiously, had been the only one on staff to receive the allegedly threatening phone calls; who had conveniently popped up at the station scarcely ten minutes after we had gone off the air; and who had taken more than an hour, he said, to locate a pay telephone in the Empire State Building from which to phone me on the night of the incident. Moreover, aside from the key held by the chief engineer (a long-time employee) and one or two others belonging to senior staff members, his was the only key to the transmitter room.

Now the young man was working alongside our chief engineer, night and day, to repair the damage. He seemed able to get right to the heart of the trouble. The police, meanwhile, put a twenty-four-hour-a-day guard around both the transmitter and WBAI's studios.

A week later, the damage was repaired and the parts replaced—thanks to a loan from another radio station broadcasting on the same frequency in another state—and WBAI returned to the air. A couple of days later the New York City Police Department, apparently convinced that no further incidents were imminent, withdrew its guards. They did promise, however, to continue their investigation.

We were not so convinced of our safety. The staff decided to stand its own guard, in shifts, at the transmitter, at least until adequate security could be installed. Detailed instructions for rapid communication were given to everyone in case anything "suspicious" should occur. On our first night without police protection I was particularly careful to give explicit instructions to a young, scatterbrained announcer who was to be on duty at the studios. He had just returned from vacation, oblivious to all that had occurred during his absence. I told him to call the person on watch at the Empire State Building if anything even slightly out-of-the-ordinary happened. Then I left the Thirty-ninth Street studios and joined Larry Josephson, who had the first shift at the transmitter room.

A few moments after my arrival the telephone rang. It was the announcer on duty at Thirty-ninth Street. The lights had flickered off a couple of times in the control room, and, while he didn't feel it was anything to be concerned about, he had decided to follow my instructions to call at the slightest provocation. I asked if anything else unusual had occurred. He said there was nothing, except that the smell of smoke from the incinerator was perhaps a bit heavier than usual.

Now I didn't wish to sound like an alarmist, but under the circumstances I did feel that this was at least worth checking out—especially since I could not, offhand, recall when the trash had last been burned at nine o'clock in the evening. Calmly, in a voice several octaves above my usual range, I suggested the announcer quickly check the incinerator.

He returned a few moments later to report that the Thirty-ninth Street basement, where the office supplies were stored for WBAI and the other organization then housed in the building, was ablaze.

I instructed the announcer to leave whatever tape was then running on the air while he called the Fire Department and evacuated the building. Josephson remained at the transmitter, prepared now for almost anything, while I hurried the few short blocks from the Empire State Building back to the studios. By the time I arrived fire trucks were pulling up. It seemed an extraordinarily rapid response, and indeed it was, for as it turned out the fire had been reported before we had even discovered it. Perhaps before it had been set.

The blaze was quickly extinguished, and even the most cursory of investigations by the fire marshals indicated arson. The lock on the front door to the East Thirty-ninth Street building had been tampered with, but the door leading to the basement had been opened with a key. The fire itself had several points of origin in the basement, and a brief inquiry with the staff present at Thirty-ninth Street at the time revealed that our Prime Suspect, the engineer, had left the building only minutes before the blaze was discovered. The Fire Department confirmed that they had received a telephoned alarm, from a pay phone, prior to the one phoned in from the station. That call corresponded almost exactly to the time the Fire Department estimated the blaze had been set.

Once again the studios were placed under twenty-four-hour-a-day police guard. Both the Police and Fire Departments believed they had enough evidence to prosecute the young engineer. A hearing date for a preliminary indictment was set for several weeks later. Though vandalizing the transmitter of a federally licensed radio station is a federal offense, no investigation was ever launched by the federal government. The New York City Police and Fire Departments both continued their investigations, gathering evidence and searching for a motive.

As the investigation proceeded, a jurisdictional dispute arose between the Police and Fire Departments. The detectives who had first responded to and investigated the transmitter vandalism felt the case was theirs, while the fire marshals who had become involved in the Thirty-ninth Street arson believed it was up to them to prosecute. The two city agencies extended only grudging cooperation to each other during the investigations, and on the date of the initial hearing the fire marshals, who had been called as material witnesses, failed to appear.

A second hearing date was set, but again the marshals, whose evidence was crucial to the case, refused to testify. The judge now set a third date, and indicated that charges would be dropped if the prosecution could not then come up with the necessary evidence. Neither the evidence nor the marshals were forthcoming, and the case, as expected, was dismissed. No further legal action was ever taken against the young engineer.

WBAI has had many moments of notoriety during its more than a dozen years as a listener-sponsored radio station, and most have been proud moments—times when principled stands have put us, for the moment, in the spotlight. Those stands have not always been popular ones, not even with WBAI's own minority audience, but they were, for the most part, acts of group conscience.

The incident I have described hardly seems to fit this category. Had our attacker turned out to have been, say, a member of the

American Nazi Party, or any right-wing extremist group, or even a lone nut with a political motive, as early newspaper accounts hinted (as opposed to a nut on the WBAI payroll), we would have garnered a good deal of sympathy from the community. But under the actual circumstances, public attitudes seemed to be less sympathetic. And though the station thrives on crises which place it in the political spotlight, this one we all could have lived without.

But rather than retch up an old unwanted skeleton simply for the sake of the tale, let us take a look at how the institution responded internally—politically and psychologically—to this bizarre turn of events.

Though any attempt at political categorization of the WBAI staff, past or present, is at best questionable—if not impossible—it can be said with a degree of safety that the majority of the staff members consider themselves stanch defenders of individual liberties, with the assumption of presumed innocence a cherished principle. So when the accused in this case suddenly was one of "our own," whose alleged crimes threatened both the institution and its people, there arose a conflict between those libertarian and humanitarian ideals and the natural instinct for self-preservation.

On the night of the Thirty-ninth Street arson, when the police had taken the young suspect in for questioning, their object, it seemed, was simply to extract a confession from him: had he or had he not committed the crime? Their methods ran from promises of leniency to threats of severe punishment to intimidation through implied, though false, claims of substantial evidence. Unable, finally, to get a confession, or even to solicit enough inconsistencies to press charges, the police were forced to release him.

Though the station manager, Frank Millspaugh, Program Director Dale Minor, and I had cooperated in the questioning, the process left us feeling uneasy, a fact we must have silently acknowledged among ourselves several times during the night. Still, we were relatively convinced that the engineer had indeed committed both the vandalism and the arson, and so it was with equal discomfort that we witnessed his release.

The conflict at the station became more intense in the days that followed, as management, not without a great deal of soul-searching, dismissed the engineer, relieved him of his keys to the studios and transmitter room, and banished him from the premises. Though he had been neither convicted nor formally accused of any crime, it seemed to be the only sane and sensible action that management, which holds ultimate responsibility for the preservation of the station, could take.

As usual, the staff demanded its say. While few disputed the suspect's probable guilt, many felt that management's action was not only precipitous and a violation of our own individual and institutional principles, but also a position counter to this country's constitutionally stated concept of justice. Had the case not involved our own self-interests, it was felt, our sympathy and support would most certainly have rested with the accused. Some staff members even pushed for the suspect's reinstatement, though in the end—and, I believe, rightly—management's self-protective instinct prevailed.

Some more speculative information about the personal motives of the people who serve WBAI might be drawn from what, at the time, was believed to be the young engineer's motive. While his behavior may have been extreme, the emotions that apparently drove him to act as we believed he did, are not unlike those that brought many of us to WBAI.

Like most of the staff, past and present, the young engineer was first a listener to the station. And, like a good number of WBAI's listeners, his attachment to the station went beyond that of most listeners to other radio stations, which are little more than a kind of audio wallpaper. Deciding to see what was behind the unusual information and peculiar personalities coming out of his radio, he volunteered his considerable technical talent, which the station put to immediate use. It did not take the newcomer long to find, and seek to become a part of, the protective WBAI "family." And, though acceptance within that family is not easily attained, it is binding when achieved. The staff is, after all, an odd cross-section of heads, as reflected in its broadcasting. Internally it is a small society set off from the one which surrounds it, though its doors remain open to those in need of its dubious sanctuary.

To become a part of this internal community, then, is the singular goal of those who have become infected with the BAI bug. It is not simply the search for employment, for the young engineer and many before and since could earn and have earned many times more than their meager WBAI salaries at jobs far less demanding and time-consuming. It is the protective, intense, and incestuous family as much as ideological identification that drives some to the almost fanatic dedication necessary to attain staff status. Sometimes this status is gained through political maneuvering, though fortunately it is more often achieved through the kind of spectacular dedication, talent, and skill displayed by the young engineer during the days immediately following the transmitter destruction—the destruction that he himself had presumably brought about in order to gain the staff's recognition and respect; it was his tribal initiation.

----

## NOTES FROM THE AUTHOR'S DIARY

SEPTEMBER 24, 1967

Paranoia. Real paranoia. Ever since that night at the police station—first I couldn't get myself to go home. Stayed one night with Millspaugh, another with Dale and Elizabeth. Now every time I come home I kick the door open first and check the place out. For what, I don't know. When I get into the car I check the engine for a bomb. Not that it would do me any good. I wouldn't know a bomb from a spark plug. . . .

He must be insane. Last week he called the cops and told them someone was firing shots into his bedroom window. Which is unlikely, since he lives on the 17th floor across from a six-story building. The police reported that the bullet holes on the wall were made by a screwdriver (which they found lying on top of a bureau, bits of plaster and all). Now there's a story going around that his former roommate was found murdered. . . . .

No wonder I'm cracking up! We should have suspected something as soon as we found out he worked for an electronic bugging company. . . .

all power to the
transmitter...

# 2

## Beyond the Quill Pen

WBAI originally belonged to Louis Schweitzer, a somewhat eccentric, socially aware millionaire, who bought it in the late 1950s to add to his collection of unusual hobbies.

(Years earlier Schweitzer had gone one day to have his hair cut by his favorite barber. It was a few minutes past five P.M. and the owner refused to admit him because union rules demanded that all work cease at five sharp. Schweitzer simply bought the shop and gave it as a gift to his regular barber, the only attached string being that he be able to get his hair cut whenever he pleased. On another occasion he purchased a New York City taxi medallion for his Mercedes limousine, so his chauffeur could earn some extra money while Schweitzer was at the office. Profits were split down the middle. Later in life, while visiting Venice, he purchased a fleet-owner's gondola and turned it over to the gondolier, so that he would have it available to him anytime he might need transportation through the canals of Venice. You know how it is with gondolas—you can never find one when you really need one. These are but a few of the legends, or myths, associated with Lou Schweitzer.)

Legend has it that Schweitzer chose this station because he could see its antenna from his residence at the Hotel Pierre. At the time of his purchase it was a commercial station, but because he was able to absorb the financial loss, Schweitzer hoped to move the programing into unusual and diversified areas. It floundered along, consistently losing money, until the New York City newspaper strike of 1960, when advertisers, suddenly unable to get their messages across in the daily printed media, turned to the city's FM radio stations. The resulting flow of income soon made WBAI a commercial success. But Schweitzer came to believe that profit and good, free radio were incompatible. He became disillusioned and bored with his commercially successful hobby, and offered it as a gift to the nonprofit Pacifica Foundation, which he knew was experimenting with two listener-sponsored radio stations in California.

Eleanor McKinney, a Pacifica pioneer, in a 1962 broadcast entitled "About Pacifica Radio," describes the offering:

*One day, while struggling with the innumerable problems besetting Pacifica in its main office in Berkeley, California, Dr. Harold Winkler (then president of Pacifica) received a long-distance call from New York. At the other end of the line was Louis Schweitzer, a remarkable man whose exceptional individuality expressed itself in his unusual philanthropies. Mr. Schweitzer said, "If Pacifica wants a station in New York, I'll give you one." . . . "I realized right then, when we were most successful commercially, that was not what I wanted at all," Mr. Schweitzer reported. "I saw that if the station ever succeeded, it would be a failure. . . ."*

Apparently it took some doing on Schweitzer's part to convince Winkler that he was not a crank, and that his offer of a completely equipped radio station, no strings attached, was for real. It is said that Schweitzer's frustration was so great that he nearly hung up the phone and withdrew the offer.

Today, more than a dozen years after its inception as a listener-sponsored station, WBAI is still considered the oddball of the electronic media. That this should be so in a land that piously parades the banner of "free speech" before its citizens is at least ironic, when "free speech" is the one absolute principle upon which Pacifica radio stations were founded, and to which they have strictly adhered. That this same government, which claims to guarantee and protect that constitutional right (not the "privilege," which the stanchest of this country's self-proclaimed defenders have mistakenly called it), should have on numerous occasions over the years denied the stations their full three-year license renewal, and brought the entire Pacifica Foundation before hearings of the Senate Internal Security Sub-Committee (SISS), opens that claim to question.

WBAI has been characterized as part of the "alternate media" or simply as "underground" radio, but the station defies such neat categorization. Most radio stations and printed journals that have worn these labels have died an early death, while WBAI and the other Pacifica stations have precariously clung to life day by day. Perhaps "free radio" would be more apt, though I'm sure most other broadcasting outlets would claim the same title, if with the merest of justification. The difference is simply that those of us who broadcast over listener-sponsored, Pacifica-owned stations say what we want to say, when and how we want to say it, as long as it is not in direct violation of Federal Communications Commission (FCC) regulations. And even then we have been known to attempt to broaden the interpretation of those regulations. All other radio and TV stations are, of course, legally permitted the same freedom WBAI enjoys, but most choose not to take advantage of it. As Schweitzer himself learned first hand, commercials and content seem to be incompatible when profit is the primary motive.

WBAI's "open mike" policy has earned it a number of different (and in some cases conflicting) reputations. Some are justifiable, others not. One example of ignorant labeling was the widespread charge of anti-Semitism leveled against the station, stemming from an incident that took place during the heated and arduous New York City teachers' strike of 1968.

The controversy surrounded Julius Lester, a black writer who, at the time, was producing a program over WBAI dealing primarily with the struggles of emerging black consciousness. He invited Leslie Campbell, a leader in Brooklyn's Oceanhill-Brownsville ghetto-area experimental school district, to be his guest one night. During the broadcast Campbell read a poem over the air that one of his eighth-grade students had written. It expressed a black youth's anger over the exploitation she felt herself to be the victim of at the hands of the community's Jewish businessmen. These

Julius Lester, expressing himself—symbolically, of course.
(PHOTO: STEVE POST)

sentiments were not subtly expressed, as the first lines of the poem indicate:

> *Hey, Jew boy, with that yarmulke on your head.*
> *You pale-faced Jew boy,*
> *I wish you were dead.*

Few complaints immediately followed the broadcast. To WBAI's regular listeners it probably seemed a natural extension of our usual broadcasting. After all, racial tensions in the city were on the point of exploding, and WBAI was playing its accustomed role in the dispute, offering an open, uncensored forum to all sides, and analytically reporting the events.

Perhaps if the program director had known in advance that Campbell intended to read the poem on the air, there would have been some prior discussion to determine if the broadcast was in direct violation of existing FCC regulations. But even that is only speculation, for WBAI does not "screen" guests before they appear for live interviews. The program director rarely knows in advance what guests will appear on what programs. The microphones are open to those with something to say, and WBAI lets them say it, in whatever form they may choose.

Often this policy makes for uniquely dull radio, but there are moments of spontaneous excellence, even brilliance. Lester's program that night was probably closer to the former, and I'm certain the poem barely penetrated the numbed consciousness of most of his listeners. In fact, I suspect that ninety-nine per cent of WBAI's listeners were totally unaware of the Campbell broadcast until several weeks later, when the United Federation of Teachers—running out of ammunition in its fight against school decentralization—began to exploit the issue of "black anti-Semitism" in the Oceanhill-Brownsville dispute. Included in their evidence was the poem and its broadcast over WBAI. The teachers union lodged a protest with the FCC, an action that gained the attention of the press, and caused what might have gone unnoticed to become a major, albeit smoke-screen, issue in a raging local controversy.

In a January 15, 1969, interview with the *New York Post*, Julius stated that the poem did not express his own view, but that such feelings could not be ignored. Additionally he charged that the UFT, in filing its complaint with the FCC, was using the station to get at Leslie Campbell. "The sad thing to me is that I feel the UFT is responsible for quite a bit of the feeling that exists among young blacks now in terms of Jews," said Lester.

And *The New York Times,* on January 21, reported on its interview with WBAI's station manager:

*Frank Millspaugh, the station's manager, said yesterday that Mr. Shanker has sought to gain support for an unpopular strike by raising the spector of anti-Semitism.*

*Mr. Millspaugh said there was anti-Semitic feeling on the part of some Negroes and some anti-Negro feeling in the Jewish community.*

*"It is the responsibility of the news media in general, and WBAI in particular," he said, "to make a full disclosure of these biases and to fully explore their depths and their origins. These feelings will not disappear by pretending they do not exist; I hope that they may be alleviated by open and public discussion."*

The Jewish Defense League picked up on all this and began pursuing Julius Lester, WBAI, and a few other lucky individuals and institutions as if we were the Second Coming of the Third Reich. On the night of Julius's next program, January 30, 1969, about two hundred pickets—members of the JDL—gathered outside of WBAI's Thirty-ninth Street studios. They were joined by a group of about twenty-five counterdemonstrators, who were there in support of WBAI's unequivocal First Amendment position. The air was tense with anger on both sides, and the presence of the press, complete with high-powered lights, camera, and sound equipment, all awaiting Lester's arrival, did little to calm the situation.

The poem, the demonstration, and its aftermath received a great deal of publicity, and charges of "anti-Semitism" and "shouting fire in a crowded theater" have continued to be leveled at the station over the years. That the latter charge is valid only if the theater is not burning has not been noted by our critics, nor has WBAI received a good citizenship award for sounding the alarm.

It was the intention of Pacifica's founders to develop a radio station that spoke to the minority. They believed that in a society which supposedly guarantees the right of all its citizens to freedom of speech, no matter what their views, there should be free and open access to the electronic media as well.

The commercialization of the electronic media, they believed, stifles those with avant-garde political and cultural ideas. Commercial radio, in order to sell its products, must cater to the lowest common denominator of intelligence. Or so it was and is believed by those in positions of power within the media. The founders of Pacifica felt it would be possible to run a truly free radio station only if it was financially supported by those who listen to it. KPFA-FM in Berkeley, the mother of us all, went on the air in 1949. Eleanor McKinney, in her broadcast "About Pacifica Radio," describes the moment when Lewis Hill, the man whose idea it all was, brought that idea to life:

*At three o'clock in the afternoon on April 15, 1949, Lew Hill stepped to a microphone, and the workmen, hammering down the carpet at the last moment, paused in their work. The rest of us were busy pounding out program copy and continuity on typewriters nearby. He announced for the first time: "This is KPFA, listener-sponsored radio in Berkeley." For a moment the typewriter copy blurred before our eyes—and the project was underway.*

KPFK-FM in Los Angeles followed ten years later, and in 1960 WBAI in New York became Pacifica's third listener-sponsored radio station. KPFT in Houston, Texas, became the fourth Pacifica station in 1969.

WBAI has managed to continue broadcasting for more than twelve years as one of this country's handful of truly free and independent radio stations primarily because of the way in which it obtains its financial support—solely through voluntary listener contributions. Listener-sponsorship has, over the years, been the source of our greatest strength: independence of funding sources which might exercise control over programing. Yet at the same time it has been the cause of our greatest weakness: the constantly precarious state of our finances.

WBAI's "subscribers" know they are vital to its survival: their continued support constitutes not only WBAI's income, but also public approval for free radio. WBAI and the three other Pacifica stations are unique in that the relationship between broadcaster and audience is one of mutual dependence. There have been times when we wished it otherwise, but for the most part it has been a satisfying relationship.

WBAI has changed with the times, yet the bulk of its programing is still aimed at a small segment of the radio audience. Issues that are today considered "safe" for discussion on network TV were being explored over Pacifica stations years before they were fashionable. Issues still considered too "hot" or controversial for the established mass media continue to be aired freely on WBAI. In two months we may hear them mentioned by Phyllis Newman on the *Tonight* show.

Because of the abundance of printed matter and the availability to the public of the electronic media, this journey up from the underground is made in a fraction of the time it would have taken in years past. Those in power within the electronic media, however, deserve little credit for this phenomenon. They have rarely knowingly aroused the conscience of the public. If it is indeed true that the public today is more aware of the way our society works (or doesn't work) than it was thirty years ago, it is hardly because the established media set out deliberately to inform them. A microphone or camera will inevitably and eventually pick up bits and pieces of the truth. Too often the truth winds up on the cutting-room floor, but fortunately enough has slipped through to give us a sketch of what's really going on.

Dale Minor, a CBS reporter and former WBAI program director and Vietnam correspondent, explains in his book, *The Information War*, his belief in the willful ignorance of the public:

*People in general want the press to tell it like they think it should be. And there undoubtedly remains in the public psyche a vestige of the reaction of societies in ancient times toward messengers who brought bad news—the stock in trade of contemporary journalism.*

*It may be that the American people themselves are growing weary* of democracy. Such a reaction to turbulent, uncertain times and seemingly insurmountable problems is not without historical precedent. If such is the case, a growing suspicion of and impatience with the institution of the free press would very likely be an early warning symptom.

I believe it is too early to make such a judgment against the American people. The public has been presented with such small doses of "reality" through the media that it hardly knows yet what to look for, much less reject. If our society survives (and the assumption is made here that it must change radically if it is to survive), it will be because information has gotten through, sometimes as a result of the media's efforts, but as often in spite of their efforts to conceal it.

Fortunately, information today gets through in the printed media, largely because of the abundance of "underground" periodicals. Ideas, as well as products, gain acceptance in the larger society by their evolution in the media. The process for a story might go something like this: the back pages of *The East Village Other*, to *Ramparts*, to *The Village Voice*, to *Playboy*, to the front cover of *Time*, and finally the ultimate in respectability—condensation by *Reader's Digest*. The final step, then, might be a TV "special."

One such example of a news story's journey to safety, from WBAI's own history, concerns the broadcast of former Special Agent Jack Levine, of the Federal Bureau of Investigation. The incident occurred in the fall of 1962, when Special Agent Levine brought his "inside" story of the FBI and its director, J. Edgar Hoover, to WBAI producer Dick Elman, after attempting unsuccessfully to bring it to public attention through other, more established newsgathering organizations. Though the McCarthy era had ended, its taste lingered on.

Chris Koch, WBAI's acting station manager at the time, describes the broadcast as "the first sustained attack on the FBI and its director, J. Edgar Hoover, ever presented by American radio or television. Indeed, at the time the program went on the air, it was one of the few attacks on the FBI available in any form."

During the week prior to the airing of the tape, Koch, Elman, and others at WBAI were warned repeatedly not to broadcast the program. The Justice Department suggested that it "would not be in the public interest." Journalists, including many who were friends and supporters of the station, advised that we would be jeopardizing our license. Numerous bomb threats were phoned in, and one staff member received a call from a labor leader in Washington claiming "inside information" that everyone at the station would be arrested within minutes of the broadcast.

But Koch, Elman, and WBAI withstood the pressure, and, backed by the slightly fearful approval of the Pacifica Board of Directors, aired the tape.

None of the more immediate prognostications of the prophets of doom came to pass. And, as usual, once WBAI had laid its institutional head on the line, without its being severed, the more conventional news media picked up the story. As Koch tells it:

*The story that Levine told was carried in newspapers throughout the country. Once Pacifica had broken it, the rest of the press was willing to use it, and several papers—The New York Times, for instance—editorialized about it. Later Fred Cook opened his study of the FBI with lengthy excerpts from the program. Meantime, we had to pay for it.*

A more ludicrous example of this process begins with an ad in the back pages of the *Los Angeles Free Press* for "flavored douche," available in a variety of flavors. The product, which was probably manufactured by a group of students in the basement of one of their homes, was sold by mail order only and delivered in the familiar "plain brown wrapper." Today it is manufactured by a large, reputable pharmaceutical firm, advertised in major women's magazines under the brand name of Cupid's Quiver, and is available, without prescription, in any drugstore. (It has been said that "there is nothing more powerful than an idea whose time has come.") When the media advertising campaign for this product gets into gear, Cupid's Quiver will be on the lips of every American.

If the media were as dedicated to communicating accurate information as they are to creating markets for useless and often dangerous products, our survival might be assured.

"We're giving them what they want," is the familiar response network vice-presidents offer to criticism of their programing. How do they know? What of significance or substance have they offered the public? If the conclusion is to be valid, the experiment must have a control, and this the media has not yet provided. Of course the public will watch and listen to moronic entertainment and bland public-affairs programing: what is the alternative?

Joseph Morgenstern, *Newsweek* columnist, writing about the removal of the laugh track from TV's *The Odd Couple,* quotes a network production chief as being "terrified at the thought of risking a good show without a laugh track, because I don't think people know how to react unless you tell them. I know it's an Orwellian concept, but people watch TV an average of five and a half or six hours a day and you really have to program them into having any reactions at all."

But the media are not entirely to blame for their own stagnation. Some credit must go to those in power in the government, to whom a truly free flow of information seems to represent a threat. Minor, in his introduction to *The Information War*, speaks of the conflict "between the democratic imperative of full disclosure and those forces and tendencies which act to constrict, control and manipulate the information the public gets."

Still, as Minor's superb award-winning reporting from Vietnam and isolated instances within the established media indicate, it is possible to gather and disseminate accurate information despite those "forces and tendencies." The Federal Communications Commission act of 1934 states that the airwaves are owned by the public, and broadcast licenses are granted "in the public convenience and necessity." The act further states, "Nothing in this act shall be understood or construed to give the commission the power of censorship over the radio communication . . . and no regulation or condition shall be promulgated or fixed by the commission which shall interfere with the right of free speech by means of radio communication."

Minor, again in *The Information War*, adds, "In point of fact, the commission has been generally careful to honor that proscription, and what censorship does occur in the electronic media (and it is massive) is both self-imposed and, in the past, has had little to do with government pressures or control."

Richard Moore, president of KQED, San Francisco's educational-television station, concurs. "No one has told us not to get into certain areas," he said in a *Newsweek* piece on public television, "but the viciousness is in self-censorship. Our network system is keyed primarily to the more conservative elements in white middle-class communities."

Former Vice-President Spiro T. Agnew's alliterative attacks on the press, President Nixon's establishment of the White House "Office of Telecommunications Policy" (OTP), and the new FCC "Office of Plans and Policy" to "develop and evaluate long-range plans and make policy recommendations in all areas of FCC responsibility and review all existing FCC policies" are, to date, the government's strongest threats against the dubious freedom of the electronic media. Yet, as of this writing, they are still threats, not legal restrictions, and rather muddled threats at that. Their intention, it seems, is intimidation rather than restriction, and if past experience is any guide, many of those in power within the electronic media will welcome that intimidation with open arms and a sigh of relief.

I suspect that, if any chose to use it, the power of even one major network, not to mention the potential power of a coalition of the three, would surely be enough to stave off an assault on their freedom even by the federal government. In fact, the networks have occasionally provided a small example of the use of this power—such as CBS's production and presentation of its investigative documentaries "The Selling of the Pentagon" and "Hunger in America" despite government pressure to abandon those broadcasts. It is, to say the least, unfortunate that such examples are so rare.

With all its limitations considered, it is possible that broadcasters have managed to convey more of the truth to more of the public in the last thirty years than the printed media have since the inception of the quill pen.

# 3 "Do Not Rush into Master" (Crawl in Slowly on Hands and Knees)

The backgrounds of those who pass through WBAI are as diverse as the patterns of pieces on a chessboard, and, as in chess, there are unifying aspects which have drawn them all to the same game.

Prior to my involvement with WBAI, I'd had a number of brief and notably undistinguished careers, including two short stints in the "checking department" (a Madison Avenue euphemism for the mailroom) of two separate advertising agencies. In fact, in the second agency I *was* the checking department.

The first company fortunate enough to retain my services had among its accounts a number of the country's largest defense contractors, a fact which left me with a somewhat uneasy feeling at the end of each day's work. So it was, at least partially, with a sense of relief that I greeted the news of my dismissal, after six weeks, from the forty-five-dollar-a-week position, which consisted of locating, clipping, copying, pasting, and the all-important "checking" of hundreds of ads placed in hundreds of different periodicals daily, weekly, biweekly, monthly, and quarterly.

Feeling as though I had firmly established myself in the field, I pounded the pavements and want ads seeking similar employment. Finally I wound up at a tiny agency that handled Manhattan apartment-rental agents almost exclusively.

Chmg 1½ rm st, slpg alc, tb in sk, bk wl, see spt

was typical of the hundreds of ads placed daily by my new employer. It was my unfortunate luck to have acquired this position during a New York City newspaper strike, when the only paper handling such ads was a hastily thrown-together tabloid called the *New York Standard*. Because it was operated by a skeleton staff, the real-estate ads were placed in no particular order, and it was my job each day to track down hundreds of such two-line ads out of the thousands of almost identical ones scattered haphazardly about the *Standard*.

As the weeks passed, my trips to the toilet became lengthier and more frequent, as did my lunch hours. Before long the boss noticed the absence of protoplasm at my desk, and my second stint in the "ad game" ended, after eight weeks.

I hit the pavements again, but at the time the job market for unskilled failures was slim. Quite accidentally, in the course of my poundings, I ran across a childhood acquaintance who had spent her formative years in a rowboat chasing nuclear submarines off the coast of New London, Connecticut, in some of the early peace activities of the Committee for Non-Violent Action (CNVA). (She had always seemed somewhat odd by neighborhood standards.) Now she was employed by a peace organization operated by the Religious Society of Friends (Quakers). She offered me work there, which I was reluctant to accept, in spite of my having taken part in a few peace protests in high school. I had not developed the sense of dedication necessary to sacrifice one's life to such noble work, though I had seen *Friendly Persuasion* (plus a star-studded stage show) at Radio City Music Hall and been impressed. Nor was I prepared to work in a dingy basement office under a bare forty-watt light bulb, typing away on a donated turn-of-the-century Remington.

But forty-five-dollar-a-week jobs were hard to come by then, even in the peace movement, and I accepted. Neither the job nor its surroundings were what I'd expected. I spent the next two years typing and reproducing leaflets and position papers in a rather plushly renovated brownstone. Though I felt far more comfortable with the people and the work than I had at either of the agencies, I finally grew disillusioned with the endless masses of words. Why, I wondered, were these people not out in the streets dressed in loincloths, leading the masses to nonviolent revolution?

One day a friend who knew I was looking for work called to tell me about two jobs: one with yet another Quaker group; the other with an offbeat listener-sponsored radio station. Since I'd had it with Quakers, I phoned the station, WBAI. (I learned, unfortunately only after I had taken the job, that the Pacifica Foundation was founded by Quakers. It's tough to shake old Friends.)

The chance to work at WBAI, in any capacity, seemed, then as now, a curiously lucky twist of fate. If I'd had any childhood dream at all it was to "be on the radio." In the early 1950s, at the misguided recommendation of a speech therapist, my father had purchased a tape recorder so that my brother, who stuttered severely, could hear his voice and speed up his recovery. But the device backfired. My brother, after listening to himself on tape, was so traumatized he could barely utter a coherent sentence for years afterward. He fled from the tape recorder in horror, never to use it again as a therapeutic device.

My father made it clear to me that I was not to use the machine as a toy, but it remained around the house to be used primarily as a device to amuse or horrify unsuspecting guests, whose candid conversations my father would record. Which is how my mother lost many of her mah-jongg companions.

For a while I complied with my father's rule, but the mystery of the forbidden machine, and my already long-standing desire to be on the radio, eventually compelled me to break the house rule.

Before long "playing radio" became my favorite solitary pastime. The half-hour "programs," done under the pseudonym "Luke Warm," consisted of the ramblings of my eight-year-old mind, combined with Arthur Godfrey–style commercials, improvised around blurbs from brand-name products found in cupboards and medicine cabinets. Later, with friends and my brother (who by this time had overcome his fear of the machine, if not his stutter), I improvised comedy sketches and mock interviews, similar in intent and structure, if not content and quality, to those I was to do years later on

WBAI with Marshall Efron, Marilyn Sokol, Barton Heyman, and other improvisational satirists.

I am not at all clear about the origins of this unusual childhood desire. I had never been an avid radio listener; my earliest media recollections are of television, the first of which entered, and soon dominated, my home and life in 1947, when I was at the tender age of three. The tube became both my surrogate friend and sibling, vying for the attention and affection of my parents. For all I knew, the real world might have been made up of three-inch-tall gray people, running around a ten-inch-square screen.

Perhaps because I could not identify with those little people, my desire to be on the radio continued on and off through the years, cropping up strongly again while I was still working with the Quakers. I decided to do something about it and even investigated the possibility of going to broadcasting school. Armed with the names of several schools, I had phoned WBAI, to which I rarely listened, but about which I had heard much, to seek information about the reputability of these much-advertised institutions. I had assumed that since WBAI was noncommercial, people there would be both helpful and unbiased.

After stating my business I was placed on HOLD by the well-modulated male voice on the other end, which returned moments later with the helpful reply, "WBAI neither supplies information about, nor recommends broadcasting schools."

"Can you suggest someplace I might call for such information?" I pressed. But he had hung up.

My second telephone communication with WBAI was a bit more satisfying, if no less perplexing. I phoned the station and spoke to Halleck Hoffman, then President of Pacifica Foundation. I asked him about the nature of the available position, a question he seemed unable to answer directly. He said there were a number of jobs open, but carefully avoided describing any of them. Finally he asked if I was familiar with bookkeeping. I told him I wasn't, recalling that I had failed the basic bookkeeping course in high school, being unable to differentiate between a debit and a credit even as the end of the term rolled around.

He asked if I would be interested in on-the-job training. "No," I replied in no uncertain terms, though I added that I would be interested in just about anything else at WBAI.

Finally he told me that WBAI couldn't afford to pay very much. I told him I knew that and would be willing to work for seventy-five dollars a week. That seemed to clinch it. He said they'd find something for me and I should show up for work two weeks from the following Monday. We hung up.

I realized then that I had neither been asked to come by for a personal interview nor filled out an application, and that, finally, I didn't know what my job was to be. In retrospect, it seems, I should have recognized the handwriting, if not the wall.

I arrived for work at 30 East Thirty-ninth Street promptly at nine A.M. on the appointed Monday, dressed in suit, tie, and shoes all purchased especially for the occasion. A plaque on the front of the run-down red-brick building read simply WBAI BUILDING. I went to the second-floor offices and knocked. There was no answer. I tried the door, but it was locked. I waited. By ten my new shoes were cutting into a rather large blister on my heel, so I slid to a seated position on the floor, the only available horizontal surface.

At ten-thirty the building was still empty, and it began to occur to me that this was either a hoax or a hallucination. Finally, at about eleven, a freckle-faced, red-headed young woman came bounding up the stairs fumbling through her purse. Her hand emerged with a key ring worthy of a warden, and as she searched for the key that would open the office door, she glanced my way. She appeared not at all perplexed by the sight of a rather heavy-set, well-dressed young man sprawled on the floor.

"Oh, hi!" she said casually and walked into the office. (I was to learn that it is not uncommon to find someone in my condition, or worse, lying around the halls, studios, and offices of WBAI.)

That was Dolores, a "between jobs" dancer who was earning a living, of sorts, as WBAI's receptionist.

I explained my presence as best I could, and she led me into the reception area, which, though not much larger than a storage closet, resembled a warehouse. There I balanced myself, precariously, on a filthy green couch supported by only two legs, both at the same end.

Dolores told me to see Margaret Jackson, who, at a few minutes past eleven, was not yet in. As I waited, slowly sliding down the length of the couch, a steady procession of assorted people entered the offices, and before long an assortment of sounds started emerging from the cubbyhole offices around the reception area. It was beginning to sound like a radio station. Finally, sometime after noon, Margaret Jackson arrived.

I introduced myself to Miss Jackson, a tall, thin, but otherwise unremarkable-looking Englishwoman. She extended her hand and greeted me with a look of nonrecognition, and then, as if suddenly recalling a bit of truly insignificant information, said, "Oh, yes, you're the new bookkeeper." Though I denied it, she stood firm.

She led me upstairs to the second of WBAI's three floors, and to a large wooden desk which held only an adding machine. From the top drawer she pulled an enormous piece of green paper.

"Now this is the daily cash-flow sheet," she said, ignoring my continued protests, which had by now escalated to desperate pleas. All to no avail. I was rapidly becoming WBAI's bookkeeper.

A couple of hours later Miss Jackson and another expressionless young woman escorted me to WBAI's bank, where we filled out all the forms necessary to make mine the authorized signature for both of WBAI's accounts. They seemed in a great hurry to transfer this responsibility, though little was known about me beyond my name.

Suddenly I was in complete control of WBAI's finances, having been employed by the organization for less than two hours. Had I known the true state of WBAI's finances, I would have realized the limitations of my new power. The station had a sum total of about forty dollars in its two accounts; the outstanding debt at the time amounted to about fifty-five thousand.

The job of bookkeeper at WBAI was not, at the time I took it over, a terribly complex one. Since there was little cash flowing, the daily cash-flow sheet required only about five minutes a day. I scarcely had occasion to open the Accounts Receivable book, and entries in Accounts Payable were made almost exclusively in one column. A couple of hours a day were spent on the telephone explaining to our creditors that I was new at the job, and that as soon as I sorted out the mess left by my predecessor I'd be sure to send them a check.

I got away with this for nine months—my entire bookkeeping career. At the end of those nine months WBAI's accountant came in to audit the books, discovered that I had lost the Accounts Payable book, swallowed his red pencil, and threatened to commit suicide unless the station manager fired me at once.

But the manager apparently viewed my loss of the Accounts Payable book as an act of charity. He relieved me of my duties as bookkeeper in exchange for a job as staff announcer. At WBAI, incompetence in the pursuit of ambition is no vice.

My replacement at the bookkeeper's desk was a seventy-nine-year-old British former banker known only as Wilson. Wilson had neither heard nor heard of WBAI, which was just as well, since he displayed deep-seated prejudices against almost everyone who was not a white seventy-nine-year-old British former banker. He was hired because the management felt we needed someone at the job who had a deep understanding of financial matters, and someone, for a change, who wasn't simply "passing through," or using the job as a stepping stone to the air. Wilson seemed the perfect candidate: he had already passed the peak of his career and was simply seeking part-time work to occupy the hours until it was time to meet that Great Auditor Upstairs.

Unfortunately, Wilson was unprepared for the condition in which I had left the books. He took to drinking heavily in the early morning, a habit which was not an aid to his work. He was already showing the signs of his age: half blind, he had attached to his desk several thousand watts of electric light, all beamed directly onto whatever piece of paper he happened to be attempting to write on at the moment.

To Wilson's mind, it seemed, almost every problem encountered during the course of a day, or during the course of history for that matter, could be attributed directly to the "colored" or the "Jew-boys." Fortunately for us, Wilson was also deaf, and unable to hear our broadcasting. Not that it would have mattered. He displayed an almost complete lack of interest in the function of the organization. Though he was with us for two years, I'm not sure he even knew we were a radio station. Wilson's sole interest was in keeping the books, which he was never quite able to master, since WBAI's books bore little resemblance to any he'd seen before.

WBAI's two control rooms and one tiny studio occupied a portion of the third floor at 30 East Thirty-ninth Street. A typed sign Scotch-taped on the door to the hallway leading to master control read DO NOT RUSH INTO MASTER. Beneath it, someone had penciled in, *Crawl in slowly on hands and knees.* Though it was not the flashing red light that I had expected of a radio studio, still it was ominous and intimidating enough to send me padding back to the relative safety of the bookkeeper's desk.

It was a full month before I was to catch a glimpse of what later would become familiar territory. The occasion was a special fund-raising event called "Live Day," during which artists who rarely had access to the mass electronic media were to perform live from our one tiny studio. The performances were followed by solicitations of funds from various recognizable staff voices.

Among the notables participating in the day's events was Charlotte Moorman, a cellist whose performances of modern compositions included the use of, among other things, a garbage pail filled with soft-drink bottles, a sandbox, various portions of her anatomy (disrobed), and a .45-caliber pistol, not to mention her cello.

Because of the limited studio space, the main problem of the day turned out to be moving the program participants smoothly in and out during the brief fund-raising announcements between acts. Somehow it became my responsibility to move Ms. Moorman, and her equipment, into the studio during one of these breaks. I'd already gotten all of her paraphernalia settled in when she asked if I'd push the sandbox slightly to the left. As I bent down to make the adjustment, the sound of a .45-caliber blank whizzed past my right ear, bouncing off the studio wall.

The pain in my head passed quickly, and faintly, above the intense ringing in my ears, I was able to hear her say, "I always test it out before a performance." I do not recall that she took such precautions with her anatomy. I sustained no permanent damage from the shot, though I listed slightly to one side for several weeks.

My own first brush with the microphones came several weeks after "Live Day." WBAI's program director, Chris Koch, an intense and intimidating little man in his late twenties who looked the part of child prodigy, stopped by the bookkeeper's desk on his way upstairs to the studios. He said he heard that I knew something about Synanon, the then young and controversial organization working with narcotics addicts. (My only previous contact with Koch had been a casual but demoralizing encounter near the receptionist's desk. As he passed by I noticed that his necktie came only halfway to his waist, a fact about which I made a lighthearted comment. He looked at me as though I were an obscene graffito on a bathroom wall, turned, and walked out.)

He asked if I would be interested in interviewing Dr. Lewis Yablonsky, author of the recently published book *The Tunnel Back: Synanon.* The offer, which seemed to come from nowhere, stunned me. I replied, "Me? Are you kidding? I'm the bookkeeper!" Koch shrugged his shoulders and continued on his way.

Recognizing that my bookkeeping career was destined to be a brief one, I reconsidered his offer after the initial shock had worn off, and agreed to do the interview.

I can't say that when I came to WBAI I didn't hope to get into the broadcasting end of it. But it was only a hope, and one that seemed distant and speculative. I had always held in awe those whose voices came out of my radio; they seemed to know what they were talking about. It has been greatly disillusioning to learn how little most of them really know; that the cultivated voice flowing about my living room has been cultivated precisely for that purpose —to create the illusion of authority.

Lewis Hill, Pacifica's founder, writes of this process in his essay "The Theory of Listener-Sponsored Radio":

Let me instance the announcer, not only to seize the simplest case, but because he will serve as the gross symbol for the writer, the musician, and all who try to make a living in the program end of radio. You will recall without difficulty, I hope, this fellow's nightly solicitude toward your internal organs. In his baritone way he makes a claim on your attention and faith which few of your closest friends would venture. I know of no better explanation of this man's relation to you, to his utterances, his job, and his industry, than one of the time-honored audition tests given to applicants for announcing jobs at certain of the networks. The test consists of three or four paragraphs minutely constructed to avoid conveying any meaning. The words are familiar, and every sentence is grammatically sound but the text is gibberish. The applicant is required to read this text in different voices, as though it meant different things: with solemnity and heavy sincerity, with lighthearted humor, and of course with "punch." If his judges award him the job and turn him loose on you, he has succeeded on account of an extraordinary skill in simulating emotions, intentions, and beliefs which he does not possess. In fact the test was especially designed to assure that nothing in the announcer's mind except the sound of his voice— no comprehension, no value, no choice, and above all no sense of responsibility—could possibly enter into what he said or what he sounded like. This is the criterion of his job.

Joe McGinniss, writing seventeen years later in The Selling of the President 1968, says, "Americans have never quite digested

1793929

television. The mystique which should fade grows stronger. We make celebrities not only of the men who cause events, but of the men who read reports of them aloud." That these men often have no idea what they are talking about has not yet penetrated the public consciousness. The electronic box is, to them, the final word in truth. Fortunately, though inadvertently, it often is.

However, Chris Koch, like most producers at WBAI, was not a typical broadcaster. With few exceptions, at WBAI over the years the best producers have not had previous professional experience. I do not mean to imply that they necessarily always know what they are talking about, but they certainly do not fit the mold of professional broadcasters. And many "professionals" would be as eager to point this out as we are.

Koch himself was partly responsible for some of the finest radio documentary work ever produced, including the Armstrong Award–winning series, *This Little Light*, about the civil-rights activities in Mississippi during the summer of 1963. Koch, and others at WBAI, have been able to produce such programs precisely because they have had little previous radio experience. For the most part, existing radio stations, both commercial and "educational," do not permit the individual producer the freedom and autonomy necessary for experimentation. Such experimentation often produces as much failure as it does success, and there is little time in traditional radio for failure. Commercial broadcasters are rigidly bound by what is felt to be the few proven successful or acceptable radio formats. They are apparently convinced that the masses are morons, incapable of accepting any sound but the most familiar, possessed of a concentration span of three minutes.

There is an unwritten rule in "professional" radio that there never be more than a few seconds of "dead air" (silence). The theory goes that Mr. and Ms. America can wait no longer to take in the next sound, and will, if left even momentarily to their own devices, change the station—the nightmare of every ratings-conscious broadcasting executive.

At WBAI, on the other hand, I have often had the feeling we could broadcast nothing *but* dead air and people would listen—as long as we labeled it "art." Larry Josephson, formerly producer and host of WBAI's morning show, believes that if we presented several hours a week of nothing but farting, the program would soon have a large and dedicated following. Audience size, even at WBAI, is rarely a correlative to quality.

My interview with Dr. Yablonsky was an unqualified disaster. I had read his book beforehand and was thoroughly familiar with the topic. Had we been sitting in my living room discussing Synanon and his book, it might have been a fascinating discussion. Instead, we spoke from a tiny room that was a maze of wires, buttons, and microphones. An engineer stood in the control room behind twin panes of thick glass, twisting dials, pushing buttons, and turning knobs on an enormous control panel that looked as though it had been dislodged from an airplane cockpit. As he started the stop clock and pointed a finger at me, I began reading questions I'd carefully typed on three-by-five index cards.

I do not know whether Yablonsky's answers in any way related to my questions—I was far more concerned with my broadcasting debut than I was with the subject. Dr. Yablonsky could have been discussing the use of acupuncture in veterinary medicine. My eyes were rigidly fixed on his mouth, and when it stopped moving I went on to the next card.

The taped interview lasted thirty minutes and was scheduled to be broadcast several nights later in conjunction with an hour-and-a-half "live" discussion with a number of ex-addicts from Synanon, moderated by Chris Koch. WBAI had recently installed a telephone "talk-back" system, enabling listeners to call in and take part in on-the-air discussions on a five-second tape delay. This was to have been one of the first uses of that system, thrown together, in the best WBAI tradition, with whatever equipment was on hand.

Disaster began to rear its ugly head early that night. About two hours prior to the broadcast Koch's voice began to give out, and by broadcast time all he could force from his beleaguered voice box was hot air. Reluctantly, he asked if I would moderate the program. I agreed, trembling at the prospect of being on-the-air "live." The program began amidst the kind of chaos typical of much of WBAI's broadcasting. I bumbled my way through the discussion, asking suitably inane questions each time the room fell silent, paying little attention to what was being said, as Koch sat helplessly by, occasionally emitting frustrated wisps of hot air.

Finally, after what seemed a lifetime, it was time to make use of the talk-back system, completely checked out and pronounced functional by our engineering staff only hours earlier. Prompted by a note from Koch, I gave out the phone number and invited listeners to join the discussion. Sure enough, the phones began to light up. As instructed, I pushed a button and said, "WBAI, you're on the air." There was no response. I tried another. Still nothing. Beyond the studio window I could see engineers darting frantically about the control room, plugging, unplugging, tripping over wires and each other, all to no avail. My panicky fingers continued pushing buttons on the telephone, while all that was heard on the air was my now high-pitched voice chanting, "WBAI, you're on the air," rejoined by feedback, line noise, and test tones.

It was clear to everyone, most of all myself, that I could not conduct a coherent conversation for another hour. Koch thought quickly (probably visualizing the end of what until this point had been an unmarred media career) and passed me a note suggesting we take listeners' questions off the air. And that is how the program was concluded, with Koch, utterly humiliated, writing questions on slips of paper and running them into the studio for me to read.

The chaos we felt in the studio apparently went unnoticed by the listeners. Radio is, in many ways, a protective medium, and, as I've noted, the mere fact of a voice mysteriously emanating from a box in the privacy of one's home creates an air of believability in the mind of the listener. Creating illusions is simple enough in radio, and the audience is eager to believe what it hears. The 1938 Orson Welles Mercury Theater broadcast of H. G. Wells' *War of the Worlds* is only the most celebrated example of such mass audience susceptibility. It is not an isolated instance. Over the years I have seen it repeated, on a smaller scale, many times.

Shortly after the 1968 student occupation at Columbia University, Paul Krassner, satirist and editor of *The Realist*, and Marshall Efron, a pudgy little man of considerable comic genius, substituted for me on one of my Saturday-night programs. Using fictitious names, they claimed to be Columbia students "liberating" WBAI's airwaves. They read all the standard station announcements, carefully followed all FCC regulations, including station breaks on the hour and half hour, and made no attempt to disguise their voices, which, after years of guest appearances on my program, were as familiar to my audience as my own.

Still, within an hour police arrived at the studios, having received reports of a student takeover and of my detention as a hostage in WBAI's bathroom. The police thoroughly searched the premises, with special attention given to the bathroom, questioned Krassner and Efron, and left, only to return again an hour later to follow up new leads. Even one of the major news wire services carried the story of the student takeover of WBAI.

On another of my programs, this one just after the 1968 Presidential election, Efron called from a pay telephone on the corner of East Thirty-ninth Street and submitted to a completely improvised on-the-air interview as Vice-President-elect Spiro Agnew. As the dialogue indicates, it was more a burlesque than a satire, and certainly not designed to deceive:

POST: *Tonight we're going to speak over the phone with Vice-President-elect Spiro T. Agnew. We feel he's been given an unfair deal in the press and tonight should have a chance to answer questions directly and tell us a little bit about what he's really like. . . . Hello, Mr. Vice-President?*

V.P.: *Uh, could you hold the line for one second?*

POST: *Sure, I'll be glad to. [A pause of considerable duration] Mr. Agnew, Mr. Vice-President . . . ?*

V.P.: *Right back here, right back, very busy fellow, very busy fellow . . . signing papers and everything. They don't make people the way they used to.*

POST: What do you mean by that?

V.P.: The first thing I discovered upon becoming Vice-President . . . well, they're not efficient, you have to tell them two or three times before they follow your directions. They don't come when you snap your fingers. They're just hanging around looking at you. I don't know, before the election they were hanging around ignoring me, now they're hanging around looking at me. It's terrible, people are just not the same as they used to be back in the good old days. [Off to one side] Give that to me, please—I said, let me have that over there will you please!? Hand it to me, hand it to me, don't just stand there . . .

POST: Mr. Agnew, Mr. Agnew, we're on the air live. . . .

V.P.: Oh! [Giggling nervously] that's just a joke. . . .

POST: I understand there is a book coming out called The Agnew Wit.

V.P.: Yes, this is a collection of some of my sayings from speeches I've made. Also off-the-cuff remarks at gatherings.

POST: I see, it must be an extremely amusing book.

V.P.: It's about forty pages. There are a lot of photographs of me and the wife. She's some dish. There are some early pictures of her.

POST: A lot of people have criticized you for what they say is your loose tongue. You've been criticized for calling a reporter a "fat Jap" . . .

V.P.: He was a fat Jap. This fat Jap reporter comes in and interrupts me in my train of thought and I told him what he was.

POST: Well, there's been a lot of criticism of you in the press about that incident.

V.P.: The press is like that. The press is looking for . . . it's trying to find means of discrediting the candidate. Those johnnys there have nothing much better to do than to try and pick a person apart. They look for, they find flaws in his personality, his appear— [There is a click]

POST: Hello, Mr. Agnew?

V.P.: Now take The New York Times, for instance: they ran a piece on me beyond description. First of all, they started to attack my friends, my business associates. They started to make reference to fine people, people who eat with a knife and fork. You know what I mean by that? I mean they eat with a knife and fork and they close their mouths when they're chewing. They're wonderful, good people, and they have children like anybody else, and the Times makes it look so their kids look at them and—[Another click]

POST: Hello, Mr. Agnew, there seems to be something wrong with the telephone. . . .

V.P.: No, there's nothing wrong with it, I'm on a pay phone right here in my office.

POST: You're calling from a pay phone?

V.P.: I have a pay phone in my office, yes. What's so funny about that?

POST: Well, I didn't say it was funny, but why would you be calling us from a pay—[Another click]

OPERATOR ON RECORDING: I beg your pardon, five cents for the next five minutes.

V.P.: Well, I'll just put another nickel in.

POST: Why would you have a pay phone in your office?

V.P.: Well, some johnnys like to come into my office, especially those johnny reporters and they're sitting on your desk shooting questions at you and the next thing you know they've got their mitts on your phone trying to call out. Now that's a nickel or a dime every phone call, so I had this here pay phone installed by my desk.

POST: Well, let me see, let me ask you this now: You've been similarly criticized for referring to people of Polish descent as Polacks. . . .

V.P.: That's right, Polacks. I used to tell Polack jokes, you know? What's the difference between a toilet and a sink, and then, you know, attribute it to the Poles, because it's my hobby. In fact, a lot of the book of my wit will be Polish jokes.

POST: You've also been criticized for saying that poor people know nothing about poverty, but that the experts know about poverty.

V.P.: That's right. Poor people are not intelligent people, that's why they're poor. Let's begin that way, shall we? If they were so smart, why aren't they rich? [Then, off to the side] Will you write that down over there please, "If they're so smart, why aren't they rich?" I want that typed up tomorrow, mimeoed, and send that to the staff.

POST: Let me ask you this: Now that you've been elected Vice-President, what do you think President-elect Nixon will have you do mainly?

V.P.: He's told me he wants the car clean—he wants that car polished within an inch of its life. Cadillac, new one.

POST: Let me ask you one final question. . . .

V.P.: You can ask me questions from now until doomsday. I'm right here in my robe and pajamas. I've got a glass of whisky. Eight ounces of solid Scotch.

POST: I didn't know that in Maryland you could buy whisky on Sundays.

V.P.: Well, I had it in the cupboard. The Missus never even saw it. I'm just going to sit here and drink until I drop.

POST: I've heard it said in the press that you don't drink.

V.P.: That's the press for you, trying to damage a person's character any way they can.

POST: As long as we're on the topic, what do you think of the young people in this country smoking pot?

V.P.: They should be shot.

POST: You're kidding!

V.P.: I think that anyone who is ruining his mind and body playing around or experimenting with dangerous drugs is a menace to society and is probably as dangerous to the American way of life as any one of your Viet Cong or your Communists there and they should be taken out and they should have their throats slit.

POST: Are you sure you wouldn't like to modify that statement a bit?

V.P.: They should be shot and then they should have their throats slit.

POST: You don't wish to modify your position on this?

V.P.: . . . then their bodies should be burnt to ashes and dumped in the . . .

POST: Mr. Agnew, this is live on radio in New York City.

v.p.: So what. This is the Vice-President-elect of the United States speaking.

POST: You realize that this will be picked up by the papers in the morning.

v.p.: Those johnnys are always out to damage a character. Anything a candidate or a politician or an elected official or an officer of the government has to say, if those johnnys want to attack him they will. They'll use anything in their greasy mitts that they can in their press, in their yellow journals, to take him apart and to damage him in front of his family and the public at large. And I tell you that nobody's safe these days. I'm going to fight back to within an inch of my life. I'm going to take them on individually, and bump their heads together.

POST: It's been said that you're only one heartbeat away from the Presidency. How do you feel about that?

v.p.: I know what you're trying to say. That given the occasion of the demise of the President, I, being the Vice-President, would then ascend to the post of the President.

POST: You put it so well.

v.p.: Scary, isn't it?

POST: Can you tell us how you would run the executive branch of the government if that sad event should take place?

v.p.: Yes. I have a friend of mine who is a registered astrologist. She's a Gemini. She's a terrific woman at a party or wherever. She's told me that she can tell me what's going to happen in the world at any given time from the way the planets and the stars are in conjunction, and what the cusp is and how the time of the month . . . And I feel the first thing we're going to do is open a cabinet seat for the Department of Astrology right there in the White House.

POST: Certainly you're not serious. You don't intend to run the country on the basis of an art or a science as uncertain as astrology, do you?

v.p.: What's your birthday?

POST: My birthday? What does that have to do . . . ?

Marshall Efron circa 1969, preparing to assault the microphone, the turntable, or me. (PHOTO: BEN SOMMERS)

v.p.: *Just tell me, what's your birthday?*

post: March twentieth.

v.p.: *March twentieth. Makes you a Leo, right? See you Leos are always questioning. You Leos and you johnnys are all alike. I don't know if you're trying to be snide, if you're going to mock me or make fun of me. This is neither here nor there. You pose a hypothetical, I answer as best I could, then you twist it around in your own cute way. You're just trying to be cute, you're trying to make fun of me. It makes no difference, I have the job, I'm not going to leave it. I have a mandate from the American public.*

post: Well, I'm sorry, I didn't mean to mock you. After all, you are the Vice-President of the United States. Is there anything else you'd like to say to the American people? You know we have a very young audience who are very much involved and concerned with the future of our country. Is there anything else you'd like to say to them tonight?

v.p.: *I'd like to tell the young people of this country to look to the new Richard Nixon. He has their best interests at heart. He is a man who is President of the United States and he will have the awesome responsibility of holding a lot of loose ends and making sure they come together. It's a terrible burden for any one man to hold and he's sensed a sacrifice on his part to maintain the government as best he can. He's a man who's thinking young and looking younger than ever and he's a virtual phenomenon and he's all right—all right in every direction. That's what I was telling the people when I was talking to them across the country. That Richard Nixon is all right, he's O.K., there's nothing wrong with him, he's all right. Richard Nixon is all right, his condition is fine, he's in fine condition, he's O.K., A-O.K. . . .*

post: Well, we're certainly glad to hear that. Let me just ask you . . .

v.p.: *I couldn't make it any clearer.*

post: No, you certainly couldn't. What message would you have for the youths who are dissatisfied with the direction in which the country is going?

v.p.: *You know, I'm glad you asked me that, because satisfaction and dissatisfaction can be lumped down there to the category of* attitude. These people who are dissatisfied, if they'd just change their attitude would find that they would be satisfied.

POST: I see.

V.P.: That's the root cause of their dissatisfaction, they have a bad attitude.

POST: Well, Spiro T. Agnew, it's been . . .

V.P.: Sperrow.

POST: Spiro.

V.P.: Sperrow.

POST: Sperrow?

V.P.: Yes.

POST: How is that spelled?

V.P.: S-p-e-r-r-o-w.

POST: Isn't this Spiro—S-p-i-r-o—T. Agnew, the Vice-President-elect of the United States?

V.P.: Yes. I'm also the Vice-President-elect of Canada . . .

POST: I see.

V.P.: . . . and Mexico.

POST: I see. But you did run with Richard M. Nixon, didn't you?

V.P.: How do you spell that?

POST: N-i-x-o-n.

V.P.: I ran with Richard M. Nicksson—N-i-c-k-s-s-o-n.

POST: There seems to be some mix-up here. Weren't you the Republican candidate for Vice-President of the United States in 1968?

V.P.: Yes.

POST: And your name is Sperrow Agnew and you ran with a man named Richard Nicksson?

V.P.: That's right.

POST: You know, everybody's been spelling your names wrong.

V.P.: That's right. That's those johnnys in the press for you.

POST: You mean this is some sort of a conspiracy?

V.P.: This is one of the ways in which they are trying to show that we cannot spell our names by constantly misspelling our names consistently. Well, we do spell our names correctly. These people are just like that. They tend to want to sell us into disgrace—to heap all kinds of vituperative remarks against us and to do things that I couldn't even tell you about on the radio, much less in person. I have received obscene letters like you've never seen before. I'm saving them. I'm holding on to them for my children until they grow up. I've received obsene pictures from Communist China. I've reported them to my postmaster.

POST: What was that you said before about being Vice-President-elect of Mexico and Canada?

V.P.: Yes.

POST: Well, would you want to clarify that for us?

V.P.: Well I'd better not say any more about that until I talk it over with my associates.

POST: Thank you very much. We're happy that a Republican administration would take so much time out to talk to us at WBAI.

V.P.: A new broom sweeps clean. Remember that, young man.

POST: Well, that's marvelous. I hope sometime you'll have more time to speak to us in depth on that subject. We want to wish you the best of luck in your new job.

V.P.: Well, Mr. Fass, it's been wonderful talking to you.

POST: No, no, no . . .

V.P.: You mean it hasn't been wonderful talking to you?

POST: No, you have the wrong person.

V.P.: Who is this?

POST: This is Steve Post, Post.

　　[Click]

Incredible as it may seem, there were numerous calls asking if I had really been speaking to Spiro Agnew: how had I gotten him to speak so candidly? listeners asked. But perhaps this says more about the credibility of Spiro Agnew than of the media.

Efron and I were so stunned by this reaction that later that same evening we put radio's credibility to what we felt might be its ultimate test:

POST: The most incredible thing just happened, but before we go into it, I'd like to give you a little background. I get a lot of odd phone calls here at the station. People call and tell me the weirdest things. When I arrived tonight at about ten o'clock, somebody said there was a call for me. I answered it and the voice on the other end claimed to be a duck. I laughed, but the party continued to maintain that it was a duck. I figured the easiest way to get rid of him was to invite him on the air.

So a couple of minutes ago the doorbell rang—I know this sounds crazy—but I went to the door and looked through the peephole. There was nobody there, so I ran out into the hall and on my way tripped over something. I looked down and there was this duck lying on its side, and there were feathers flying around. The duck— and I know you're going to think I've lost my mind—began talking to me between gasps for air. So the duck is here in the studio now. I must admit I feel silly and I don't really know where to start. I've never interviewed a duck before. First of all, what's your name?

DUCK: John. John Duck, or I might say John Drake.

POST: John Drake. And you're a regular duck?

DUCK: I am an adult, full-fledged duck. I'm sorry about the chair . . . I didn't know . . . I befouled it. Ducks will be ducks, won't they?

POST: How did you become a talking duck?

DUCK: Well, I don't know. I was always different as a duckling. People said I was homely, so I used to spend a lot of time by myself. I listened and the next thing you know I just went over to a fellow and I said, "Could you splash some water on my feathers, it's hot." He ran down the hill and was gone and I had to find a pond.

POST: Are you sure you're not putting me on? Let me just pull some feathers and see if they are real.

DUCK: Ouch! Would you please stop that, you're hurting me. I just happen to own a small hardware-supply store in upstate New York and someone said you might be interested in going to New York City and make a fortune as a talking duck and I said, "Well, I'll try it," and that's why I gave you the call.

POST: Did your parents talk?

DUCK: No. They were regular ducks. I never knew my father.

POST: Oh, I'm sorry.

DUCK: Well, that's the way ducks are: you just can't tell who's your father. They do look alike.

POST: *I see. I thought you meant your father passed away when you were young.*

DUCK: *I don't know whether he did or not. He might have been on someone's dinner. Long Island is known for its duck population.*

POST: *Do you also quack?*

DUCK: *No, I don't.*

POST: *You don't even imitate a duck quacking?*

DUCK: *Quack, quack. That's about as close as I can get, but it's about as bad as anybody else's quack. No, I'm just a regular homebody. I don't do anything different than anyone else. I just run my hardware-supply store and mind my own business.*

POST: *I just can't get over it, a talking duck. That must be terribly uncommon.*

DUCK: *I have a wife.*

POST: *Is she also a talking duck?*

DUCK: *No. She's a woman. She drove me down here. You don't think a duck can drive, do you? She's double-parked waiting for me to get through with this interview.*

POST: *Do you swim a lot?*

DUCK: *Oh yes, I put in an hour or two every day in the pond.*

POST: [Unable to say anything, giggling]

DUCK: *I wish you wouldn't laugh at me.*

POST: *I'm sorry, I just don't run into that many ducks that talk and it's such an odd sensation. . . .*

DUCK: *I have a friend named Charlie Chicken. He does a lot of talking too. But he's screw-loose, if you know what I mean. Chickens are addlepated. He talks all the time—says nothing—nothing worth listening to. He's just a talking chicken . . . a rooster, I mean. Charlie Chicken—lives in Greenpoint, Long Island.*

POST: *Certainly you would have to be called a member of a minority group, a talking chicken . . .*

DUCK: *I'm a duck, a talking duck.*

POST: *I'm sorry, a talking duck. Do you find you're the subject of much prejudice?*

DUCK: *I'm the only duck in my neighborhood. Now, if another duck family were to move in, I'm sure that I would feel some static from my neighbors. But as I'm considered an exotic—physically—I* don't feel any prejudice. I belong to the club. I peck a little at the golf clubs and I'm very good in the pool.

POST: How does the community feel about your mixed marriage?

DUCK: She was considered a homely girl, and they figured that if no one took her at least I did. I provide a home for her, and we're happy.

POST: Well, fantastic as it may seem, folks, we've been talking with John the Duck. Is there anything else you'd like to say?

DUCK: I'm thinking that for Thanksgiving a lot more people could be better off eating roast beef.

POST: Thank you, John, and I want to wish you the best.

DUCK: Thank you.

No one called to ask if I had really been talking to a duck. There are limits to the power of the media, and the susceptibility of its audience.

The Talking Duck...

# 4

# "Mumble It with Authority"

Driving into New York City from upstate some time ago, I had my car radio tuned to a local, commercial FM station which, at the time, programed several hours of "progressive" rock each night. ("Progressive" is a term the music industry applies to anything with a heavy beat that runs much more than three minutes fourteen seconds. Other standard designations include "top-forty" —anything with a beat that runs less than three minutes fourteen seconds, or is among the forty most popular "hits" on the charts, or does not interrupt the smooth flow of commercial messages; "middle-of-the-road"— Frank Sinatra, Kate Smith, Jack Jones, etc.; "easy listening"— Mantovani, Jackie Gleason and orchestra, and the Boston Pops doing a medley of almost anything; "light classical"—Ravel's "Bolero," the finale of the *William Tell* Overture, or the Boston Pops doing anything *but* a medley; "classical"; "country-and-western"; and "oldies"—anything pre-Dylan or the Beatles, or what used to be called simply "rock-and-roll.")

Soon the sounds of progressive rock began fading in and out, integrating with the oldies. I assumed at first that the station was doing a "mix" of old rock and contemporary rock to illustrate the evolution of that form of music. For almost an hour the station mixed Bob Dylan with Elvis Presley, the Beatles with Bill Haley and the Comets, the Incredible String Band with Frankie Lyman and the Teenagers.

It was a marvelous hour of radio, and one that I was surprised to hear on a commercial radio station, since, almost without exception, such stations stick rigidly to their chosen formats. Creative deviations, so the theory goes, are not conducive to the ends of commercial radio—finding and holding an audience, hour after hour, keeping the sound standardized and easily recognizable—all in the end for the sake of sales. Since more creative endeavors might render the station unrecognizable, as commercial programers believe, the regular listener might pass by his favorite frequency, find something more fulfilling, and never tune back.

As I got closer to the city, I recognized the creative force behind the splendid hour of radio: a wire coathanger substituting for my VW's missing radio antenna. The mix was created by two stations broadcasting on the same frequency, one upstate and another in New York City, which faded in and out of each other's signal area—the New York City station sticking to its format of "Golden Oldies," while the upstate station pursued its progressive format. The creativity, in this case, was arbitrary and accidental, the product of the media's technological inadequacies.

I had neither heard nor heard of a "mix" before listening to WBAI. My first exposure to this form of radio came through Bob Fass, who produced *Radio Unnameable*, an all-night "free-form" program for more than a decade. It is Fass, in fact, who is credited with the major contribution toward the development of this form of radio, though others at Pacifica laid much of the early groundwork for his experiment. But it is Fass who is the undisputed "master" of the spontaneous mix—working live, blending sounds from several different sources, and creating with them a new sound, or one whose meaning differs from the sum of its parts.

It was Fass, and his kind of free-form, spontaneous radio that most struck me during my early days with WBAI, and it was Bob who took me under his ample wing and taught me something about radio—or untaught me what I had learned in a brief radio course at New York University. I was so influenced by Fass, in fact, that for my first two years on the air I was consistently mistaken for him.

I first met Bob Fass in the winter of 1965. He had just returned to the station after an enforced absence as a casualty of one of the early political struggles that plagued WBAI during its infant years. Now Fass was rejoining the staff as an announcer, and he would resume his late-night program, *Radio Unnameable,* on a weekly basis.

I rang the switchboard one day to locate this new fellow Fass so I could get him to sign the forms necessary for the bookkeeper in me to issue him a paycheck. Dolores said Fass was sitting beside her at the receptionist's desk. And indeed, there he was: a gigantic man with receding blond hair and thick black-rimmed glasses, with hands so huge they appeared to dominate his enormous frame. His voice, soft and gentle, which I'd heard coming from the office monitors, seemed somehow detached from his body.

I introduced myself and asked him to fill in the forms. As I got up to leave I commented, only half seriously, that based on the observations I'd made during my first four weeks of employment with WBAI, his job seemed the "softest." He replied that if I thought so I ought to come up to the control room when I was free and watch what he was doing.

Whether or not I had calculated the comment to draw that response, it nevertheless could not have been more perfect had I written the dialogue. So, two hours after our first meeting, I was in the control room watching Fass do station breaks, introduce programs, cue records, splice tape, push buttons, turn knobs, switch switches, read meters, keep logs, and generally fumble his way through what I later learned was called a "board shift."

The "board," or console, is the central nervous system of a radio-station control room. It determines which program source (turntable, tape deck, cartridge machine, microphone, studio, etc.) is to go out over the air, and at what level, or volume. At WBAI, and most small stations, the engineer operating the board and the announcer are the same person. But that is where the similarity between the duties of a "combo man" at WBAI and other small radio stations ends. There is little in the way of a set format at WBAI, and almost no written copy. Programs run arbitrary lengths, as much as a half hour long or short. It is the duty of the announcer to make sense of and provide continuity between and around the dozens of tape boxes and records that are thrown at him, usually moments before (sometimes moments after) their scheduled broadcast time. He or she must also choose music, improvise on-the-air fund-raising pitches, and fulfill the legally required technical paperwork. In the end, the board operator is the final production center, covering up, and in some cases pointing up, the often gross production mistakes, miscalculations, or oversights. It is a demanding job, and one that is frequently underestimated or misunderstood by the rest of the staff, who basically believe that the announcer sits on his ass all day, occasionally identifying the station.

I spent every possible moment of the next two months in the control room with Fass. If his shifts were during office hours, I simply ignored the books (probably my greatest contribution to the job); if he was there at night, I stayed. He taught me the basics of running the equipment, tape-editing, and talking on the air, which consisted of two brief but pointed pieces of advice: speak into the microphone as though you're talking to one close friend; and, if you can't pronounce it right, mumble it—with authority. They have been of equal value.

It is virtually impossible to learn to be a staff announcer at WBAI without doing it—there is no place to practice. Training at other radio stations usually proves quite useless, if not a handicap. WBAI does things its own way, and that way bears little resemblance to the ways of other radio stations. So, after Fass had taught me all he could, he suggested I talk to George, the chief announcer, about filling in during the summer for vacationing announcers.

I had spoken to George only a couple of times. He was an imposing-looking fellow. Middle-aged, with a crew cut, he was at that time virtually the only person at WBAI who came to work in a suit. He had a booming bass voice that bounced off the walls before it reached your head. Geo. was from the old school of radio, the Milton Cross era. He had retired from radio ten years earlier after a notably dull career, sold insurance and mutual funds, and now was working his way back into the business at WBAI. He was, to put it mildly, out of place; six months after his arrival he was back selling insurance and mutual funds.

In his dealings with me as bookkeeper, George had been efficient and businesslike, qualities infrequently encountered at WBAI. I had no idea that he had joined the staff only days before I had, and, like myself, was still bewildered and bumbling his way about. Several days passed while I screwed up my courage to talk to him. Then one day he rang the bookkeeper's desk and asked me to come up to his office. Fass, apparently sensitive to the fact that I was too unsure of myself to approach him, had done it for me.

As it happened, George had been the announcer on duty the night of the ill-fated Synanon program. Still new to the station, he had assumed that I was an old hand, and that assumption—along with Fass's recommendation—left George with the impression that I could easily handle the job. All that was needed now was a routine audition tape to be passed along for the manager's approval.

The following afternoon I was placed in the studio with several pages of copy, including introductions to concerts containing what seemed to me unpronounceable Italian, French, German, and Russian names. I did my best, which was not much, and when I had finished reading I looked through the studio window at George. He was visibly shaken. "Let's try that one more time," he said.

Three days and some two dozen tries later, he had spliced together an acceptable audition tape. The tape was copied, so the splices would not be detectable, then passed along to the manager, who approved my hiring a couple of days later, a decision he more than likely had subsequent occasion to regret.

Finally, on a Saturday late in April 1965, barely six months after I had unwittingly stumbled into the cash-flow sheet, I made my announcing debut on WBAI, garbling beyond recognition the names of a group of Russian classical composers. When I had finished the closing announcement of the concert, the telephone rang in the control room. George answered it and handed me the receiver. On the other end was an irate woman with a thick Russian accent, demanding to know why WBAI had permitted an uneducated idiot on its airwaves to demean the memory of such highly esteemed Russian artists. She would cancel her subscription, she said, and send a firm letter of complaint to the manager.

I was crushed. I begged her to forgive me, explained that I was a novice, and firmly made up my mind to forgo that glamorous career in broadcasting, which, only a short moment before, had seemed within my grasp. All that remained was to determine which method of suicide would be the least painful. Suddenly the voice on the phone lost its Russian accent, began to giggle, and revealed itself to be our not-altogether-amusing receptionist, Dolores.

By autumn 1965, a full-time announcing position had become available. I was hired to fill the vacancy, not entirely because anyone thought I would make a good announcer—nobody was certain about that. What they did know was that I was a complete failure as bookkeeper. Thus arose one of those recurring institutional conflicts between its intended devotion to the humane treatment of its people and the efficient operation of a radio station.

The process of becoming a staff member at WBAI is usually a long and tedious one, occasionally boiling down to a question of endurance under extreme duress. One's qualifications for any position are scrutinized, of course, but then there is the subtle, undefined process by which the existing staff sniffs out a newcomer, a politicized and less-structured version of fraternity initiation. The term "BAI person," or "BAI type," is often heard around the station, yet I encountered no one who could define the terms. The staff, like a pack of wild dogs, works on instinct. If the newcomer's scent proves satisfactory, he or she is permitted to run with the pack, and there is little that can be done, short of a mass uprising by a segment of the staff, to eliminate that person from the pack. Social acceptance by a segment of the staff is the closest one can hope to come to job security.

Although there are often strong differences of opinion among staff members, and heated debates—even fights—occur daily, the feeling of protective community generally prevails. Larry Lee, cofounder and former manager of Pacifica's Houston station KPFT, describes Pacifica as "the largest and most successful utopian anarchist community in the United States." Yet Larry Josephson says, "There are no two people at Pacifica who can stand each other's guts." There is evident truth in these conflicting statements;

the fact that these two men have found contented employment in the organization each describes so differently provides some insight into the schizophrenic nature of the institution.

Perhaps the clash of ideologies was most simply summed up by Bill Schechner of KPFA, Pacifica's Berkeley station, and a former WBAI producer, who began his response to an interviewer's question this way: "We at the station . . . uh . . . that is, I should say 'I'—'we' is a dangerous word to use around WBAI . . ."

# 5

# A Failure to Communicate

*Steve Post is completely bald.*

*It is quite easy to slip past the two sleeping great danes and the stoned Pinkerton guarding the entrance to WBAI's remodeled Church on a Sunday evening at two* A.M.

*I open the inadvertently unlatched, foot-thick iron door and make my way through the maze of underground tunnels, passageways, stairwells, and catacombs that lead to master control. The hallways are empty except for a few rusted bicycles, a mattress, an electric range, a live stallion, a Volkswagen Micro-Bus, and a middle-aged nun who paces back and forth with her hands clasped under her chin muttering, "What have they done? What have they done?"*

*I slowly push open the door to master control.*

*Steve Post is seated at the console, totally bald, in a very gray business suit and tie, running the fingers of his left hand over hundreds of dials, knobs, and switches. He is telling jokes to a blinking light bulb.*

*"A chicken is standing on a street corner—"*

*He sees me, crouches under the console and shouts, "Who are you—what do you want—how did you get in here—I don't have any money—Power to the People—Off the Pigs—Right on—please—please—my mother is very sick—"*

*"Hello Steve. I'm Ira Epstein."*

*Averting my eyes, he stammers, "So you're Ira Epstein?" He* hobbles over to a far corner of the room and grabs a thick wooden cane.

"I only use it on damp nights," he says, downing three or four small blue pills.

I offer my hand and meet with an embarrassed silence as I notice his seemingly empty right sleeve, with a shiny hook poking out beneath his cuff. He stumbles and offers me his left hand. It is clammy and cold and several fingers are missing.

"I'm pleased to meet you," I say.

"You'll have to stand on this side if you want to speak to me," he says, pointing to his left. "Deaf in one ear, you know."

"That's too bad," I reply, my stomach sinking.

"Not really. I save money by buying Monaural. Ha, ha," he says, wiping his nose against his left sleeve and staring at the ceiling.

I notice his left eye staring down at my shoes while the other meets my gaze. He is aware of my staring and says, "I'm sorry, it hasn't been adjusted yet."

"Oh." Wiping a tear from my eyes I point to the blinking lights on the control panel and say, "This is all very impressive."

After a five-minute pause (during which I amuse myself by watching a husky rat cart off a huge, discarded piece of head cheese, and then, several minutes later, return it uneaten) Steve replies, "Well, it's no Pablo Light Show, if that's what you mean."

He continues to stare at my shoes.

Suddenly the room is very warm. I am sweating.

"I notice you have a rosary on your desk over there. Yet you give the impression over the air that you're Jewish."

He wheels his chair around so that his back is facing me and speaks in measured tones. "I usually don't like to have people in here when I'm on the air."

"Well, I'm doing this piece and I need a little more information."

He reluctantly agrees to meet me at his apartment several days later.

"Well, I won't keep you from the enema lady a minute more than is necessary. Thank you very much for seeing me—" Too late I realize my dreadful faux pas. He puts his good hand over his bad eye and says, "That's quite all right; out of sight, actually. Ha. Ha. You understand."

"Sure."

As he closes the door behind me, I am seized with a perverse curiosity and peer back into the control room through a tiny window as he prepares to go on the air. He takes his cane and spears a long-haired toupee which has been crawling slowly across the floor. He fastens the toupee to his head with a rubber band under his chin, and removes a set of dentures from a glass of water in a drawer. Slipping them into his mouth, he peers into a nearby mirror for several minutes, smiling, kisses the rosary and holds it tightly to his chest, clears his throat, and says, "Well, we're back."

On the appointed evening at the appointed hour I approach Post's Upper West Side apartment building. Old men are seated near the curb playing cards on folding tables, while fifteen or twenty old women sit on the stoop and talk, blocking my entrance.

"Excuse me," I say.

"Who you wanna see?" asks one old lady.

"Steve Post," I reply, wiping my forehead of a baked apple thrown from an upstairs window.

"You wanna see Steven. Stevela is a goot boy. You want I should show you his 'potment?"

I edge my way up the stairs amidst a sea of flabby breasts and buzz for Post. Minutes later he appears, wearing a pink smoking jacket and tan slippers, white socks rolled down beneath his ankles, and leads me into his apartment, tastefully furnished in Fortunoff modern.

Post seems more talkative and relaxed than he was at the studio. The ⅔ empty jug of Manischewitz Extra Heavy Double Sweet Malaga he carries into the apartment may have helped.

The apartment is spotless.

Post's collection of Mantovani records resides in a cabinet lining the entrance hall, each album categorized and placed in a labeled compartment—semi-classical, semi-semi-classical, light-semi-classical, moog-synthesizer-semi-classical and "for lovers only."

"Please wipe your feet; or better still take your shoes off," he says, straightening one of the plastic floormats that lead to each room.

*The sofa is covered with plastic. So is the coffee table, the phonograph, and the vase of artificial flowers that stands near the window.*

"It keeps the dust off. You know what I mean?" he says. "Please sit down," he adds, pointing to a folding chair facing a blank wall.

A bulletin board hangs on one of Post's living-room walls containing mementos of his career at WBAI: a picture of himself at the console and an autographed 8 × 10 of Dustin Hoffman. Post takes down Hoffman's photo, places it three inches from my left eye, and moves an ash tray onto my knee.

"When he came to WBAI to read from War and Peace I rushed up to him and he gave me his autograph. It was so exciting. He knows Judy Collins, you know." He glances at his watch. "Excuse me, I've got to take my medication."

He gathers several varieties of pills and capsules, washing them down with sips of Manischewitz. I help him tie the rubber tubing around his arm ("diabetic") as he picks a thread off the carpet.

"Please excuse the mess."


He grew up in New York City.

"I'll never forget one day when I was eight years old, March 19, 1951," he remembers. "I weighed two hundred and forty pounds. I was three feet tall and my stockinged feet weighed sixty-five pounds each. I was so fat that no one would be my friend and I would cry a lot. I try to compensate for that now by ignoring everybody and being mean."

Post worked briefly as head bookkeeper of WBAI before attaining his present post. "I thought I'd have that job forever, until a staff member told me, frankly, that my physical make-up just didn't suit the rigors of the job. I was too fat—and too dumb."

Post worked at WBAI in 1962 and then went on to work at WBAI in 1963. In 1964 Post worked at WBAI and soon after he spent 1965 at WBAI. His big break came in 1966 when he worked at WBAI, leading to one of his best seasons at WBAI in 1967. He moved up to WBAI in 1968, and, to the surprise of none of his friends at BAI, continued his upward climb to work at WBAI in 1969. He appeared at WBAI in 1970 and after a brief bit of moonlighting on the college circuit, he hit WBAI in 1971. Currently Post is doing a stint on WBAI and expects to make 1972 his biggest year yet, possibly trying for more personal appearances on WBAI while taping spots for WBAI and appearing live on WBAI.

As he puts it, "I don't want to run the risk of getting bogged down."

Post is sitting on a plastic floormat in the middle of his living room guzzling wine as spit drips down his chin.

I clear my throat. "I'd like to ask you about recent charges leveled against you by several staff members and listeners stating that you are insensitive to the needs and oppressions of Blacks, Women, Homosexuals, American Indians, Puerto Ricans, Chicanos, Cubans, Political Prisoners, and the Criminally Insane."

"I wouldn't want my sister to marry one—ha, ha," he replies, his head bobbing back and forth, his glazed eye rolling into his head.

I try another. "Several years ago you were strongly critical of marijuana. What are your current feelings toward soft and hard drugs?"

"I—I—I—" he replies, his head falling to the carpet with a thunk.

"Are you feeling all right?" I ask.

But he doesn't answer.

He is lying on the floor now, his thumb in his mouth, his knees tucked under his chin.

I move a plastic floormat onto his sleeping body, kiss him on the forehead, and let myself out.

Ira Epstein
BROOKLYN, N.Y.

As soon as Wilson had sorted out what remained of the station's financial records, and I had become familiar with the announcing routine, I was free to devote myself to the task of learning about programing. This was no simple task. Joanne Grant, formerly

WBAI's news director, once said that the typical Pacifica training course consisted of directions to the john. In most cases, even that is learned by trial and error.

One day, during my first weeks at the station, I was called to the manager's office for a review of the daily cash-flow sheet. When I arrived, there was a conversation in progress among the manager, the production director, and a recording engineer about finding someone to do a remote that night. I remained outside the conversation, waiting to do my business with the boss, paying only half-attention to what was being said. Before long I noticed a hush had fallen over the room. When I looked up, six eyes were glaring at me and three mouths were smirking.

"How would you like to do a remote tonight?" one of them asked. "I'd love to," I said. "What's a remote?"

A remote, they explained, is simply taking a piece of portable recording equipment to a location outside the station to record an event for later broadcast.

I told them I'd be willing to do it, but I had no idea how to operate the equipment.

"No problem," said one, with only a hint of reservation. And within fifteen minutes I'd had a crash course in how to work the Nagra, the finest piece of portable recording equipment available at the time—and the most expensive, worth, along with the microphone, cables, and related equipment, nearly two thousand dollars. My last experience working recording equipment had been my father's Webcor, circa 1955.

I arrived on location that evening at an elegantly draped old hall only a few blocks from the station, and set up the equipment. When I plugged in the machine the entire electrical system of the old building blew out. I had not been forewarned that the elegant hall was one of the few remaining structures in New York City that had not, in the year 1965, converted from direct to alternating current.

No matter. It was important enough, in the view of somebody at the station, to have the contents of that particular lecture/debate/symposium/workshop on tape and available to the audience to risk destroying two-thirds of the station's remote recording equipment. It is a typically noble, typically foolhardy method of operation, still the standard of the station today, despite the recent proliferation of requisition forms in triplicate.

Because each staff member is totally immersed in his or her own work—usually amounting to the work of several people—there is little time left for training newcomers. There are no secretaries and no assistants. Each producer, in addition to the creative work of producing programs, must handle all telephone calls and correspondence relating to his or her area of programing and listen to hundreds of hours of solicited and unsolicited tapes. These might include a girl from Queens reading her own sonnets of unrequited love over a scratchy record of Ravi Shankar playing an evening raga at the wrong speed; the comments of the leader of a new organization dedicated to overthrowing the state by meditation; a three-hour lecture on the political and philosophical implications of binary fission, or vegetarianism, or a discourse on the mating habits of tree toads. It is boring, oppressive, and sometimes depressing work, but it must be done. Occasionally a creative masterpiece will be uncovered. Very, very occasionally. But it is part of the obligation of a Pacifica station to listen to all voices within the community, no matter how dull they may be.

Chris Koch, who has worked with Pacifica stations in San Francisco, Los Angeles, and New York as news director, public-affairs director, and program director, describes the situation in his essay "On Working at Pacifica":

The rooms are piled high with old copies of the New York Times, dozens of stacks of magazines (some well known and national and some obscure mimeographed sheets), and odd-shaped boxes of tape. If you took the time to look through these tapes you might find a box from North Africa with a note attached to it with a rubber band saying something like this: "I had a chance to interview Ben Balthazar on my office dictaphone. The quality isn't too good, but this is one of the most inaccessible guerrilla leaders in Africa today." More frequently these unsolicited tapes are less exotic. "Attached is a tape recording of my thoughts on the graduated income tax. I have been systematically excluded from other radio stations, but I am told that you still believe in free speech." Nine times out of ten, such tapes are completely unintelligible. Some staff member has to listen to it, write a note, and mail it back. But once in a while it just may contain something significant. That, in a sense, has been the story of Pacifica.

Most of my work during the five years I spent with Pacifica . . . was routine (as the work of most paid staff members is). We audition tapes, answer letters from pleased and irate listeners, and try to get some of the innumerable program ideas recorded. We argue on the phone or in the reception office with great numbers of people who seem destined to be prosecuted and denied their rights. "Do you know, Mr. Koch, the FBI has been sending radiation through my walls because of my criticisms of the Catholic church?" Or, much more frequently, we are threatened, "You recently broadcast a commentary by the Socialist World Revolutionary Council. In the interests of equal time, we demand that you play our rebuttal as representatives of the Socialist Workers Classic Party. If you refuse we plan to file a complaint with the FCC."

. . . The average staff member is harassed by innumerable details. Not only is he unable to produce half the programs he wants, but 75 per cent of those he can do are second-rate. He knows this far better than the audience.

. . . Most Pacifica stations broadcast something like nineteen hours a day. About half of this time is devoted to talk programs. That amounts to the preparation and broadcast of a 270-page manuscript every day, or around 80,000 words. An hour interview takes from 10 to 20 hours to prepare, record, and complete for broadcast. An hour talk takes from 20 to 40 hours, and a good documentary may take anywhere from 60 to 240 hours to do well.*

And so, to a large extent, it is up to the newcomer to figure it all out for himself. If you ask, and continue to ask, you may be shown how to splice two pieces of tape together. But the pressure of pro-

* From Eleanor McKinney, ed., *The Exacting Ear* (New York: Pantheon, 1966).

ducing mostly original programing almost twenty-four hours a day is immense, and the staff has little time left over for teaching the basics of interviewing or documentary production, or any of the other facets of Pacifica's programing.

WBAI's programs are, for the most part, the creations of individuals. Producers are not given assignments, except when covering stories for the day's news, and they are free to do programs of interest to themselves, or that they think will be of interest to some segment of the audience. The producer does the interviews and research, makes the contacts, gathers all material, edits the tapes, writes the script, chooses music, acts as the narrator, and even does most of the engineering himself. Clearly, there is tremendous opportunity for individual style and innovation in producing most kinds of radio programs, but there are also great numbers of skills and rules that must be learned and observed, first by quietly watching over others' shoulders, then by trial and error.

My own first major attempt at production provided me with much to learn from—if little else. I was glancing through WBAI's December 1965 program guide and came across a program entitled "Christmas 1965: Sounds of New York," scheduled for Christmas Day. It was described as a collage of sounds of the season, to be produced by Dale Minor. It seemed a light, nonpolitical program, a good place for me to begin. Also, I knew Minor had recently assumed the position of program director and would probably be far behind schedule, so I asked if I could work with him on the program.

My proposition was met with a blank stare. I was right. Minor was far behind schedule—so far behind, in fact, that he could not even recall scheduling the program. He looked at the program guide, remembered, muttered a string of obscenities, and sent a young volunteer upstairs with me to be outfitted with several thousand dollars' worth of portable recording equipment, which neither of us had the remotest idea how to operate. We were given a ten-minute crash training course, and sent to roam about the streets of New York—two bewildered producers in search of a radio program.

For three days we wandered through stores, rode subways and busses, walked the streets recording Salvation Army bands, and interviewing sidewalk Santas, whose "ho-ho-ho"'s were flammable. Once we jumped into a cab to return to the station with tapes recorded at Rockefeller Center, turned on the concealed machine, and tried to coax the driver into becoming part of our ill-fated production. The conversation went something like this:

POST: *Do you notice any difference in people's attitudes around this time of year?*

DRIVER: Whaddya mean?

POST: *I mean are people any nicer or friendlier around Christmastime?*

DRIVER: No.

POST: *So you think the Christmas spirit is a myth?*

DRIVER: *Listen, mister, I mind my own business. I don't talk to no one. Somebody gets into my cab and gives me trouble, I ignore him. It don't bother me. I figure all I gotta do is get him where he wants to go. Then maybe when he gets outta the cab he'll break his leg. Fuck him.*

In a few words it seemed he had summed up how Man deals with the frustrations of urban life. Had the batteries not failed, it would have been a cherished piece of tape.

In the end we returned to the station with about twenty hours of raw material. In the three days remaining before the scheduled broadcast date, we had only to edit those twenty-odd hours to one interesting one. But we had already made enough fundamental errors to prevent that from ever happening. Not only had the equipment failed during a number of crucial segments, but we had also neglected to note the contents of the thirty-or-so reels of tape we now had to work with. That meant that each time a specific few seconds was needed we had to go through each individual reel. By Christmas Eve we had put together twelve minutes of the program. My volunteer coproducer had decided that the lucrative family business wasn't so bad after all. By two in the morning I had fallen asleep with my head on a splicing block, listening for the hundredth time to a Salvation Army band's all-accordion rendition of "Joy to the World."

During the early and mid-1960s, WBAI's programing was aimed at a rather small, elite segment of the local community. The staff was made up primarily of those who had gained their experience at the mother station, KPFA in Berkeley, whose programing at times had given the impression that it was a cultural and political arm of the university. And so WBAI's programing, during its formative years, followed in this tradition.

The programing schedule included hours of lectures and symposia on everything from art to zoology; endless political programs ranging from the far right to the far left, with heavy emphasis on viewpoints left of center; a good deal of music and discussion of music, especially classical, folk, ethnic, and jazz works by composers frequently ignored by most of the media; and discussions, presentations, readings, and criticism of plays, movies, books, poetry, and dance, again with heavy emphasis on the slightly obscure and avant-garde. The "typical" WBAI listener, then, might have been described as middle-aged, well-educated, politically left (possibly a 1930s union activist and/or veteran of the Lincoln Brigade), and culturally and artistically sophisticated. I certainly would not have been a listener during those years. In fact, I knew no one who listened, or, at least, no one who would admit to it.

There were, of course, other aspects to WBAI's programing: occasional satire, usually politically oriented; in-depth, if somewhat incomplete, newscasts; occasionally brilliant news or public-affairs documentaries; and once in a while a fine sound collage or preproduced mix. But programing was predominantly informational; and though many staff voices were heard on the air, they remained relatively distant from the listener. Their function was to present data, or to draw it from other sources, but rarely to express opinions or share experience, or communicate with the listener as an equal, on a personal level.

The only notable exception to this programing format at the time was, as I have noted, Bob Fass's *Radio Unnameable*. Besides working a full schedule as a staff announcer, Fass was producing and acting as host of *Radio Unnameable* on Friday nights, beginning at midnight, or whenever the last scheduled program concluded.

There were no rules and no boundaries for *Radio Unnameable*, as its name implies. One night Fass might play hours of uninterrupted music—all kinds of music—picked at random or relating to a theme. On another night he might conduct an interview with a Movement person, musician, artist, film maker, Buddhist monk, organic gardener, thief, or just about anybody of interest who might walk in off the street with no other broadcast-medium outlet for his message. Another night might be devoted to phone calls from listeners, again thematic or random, or just one phone call for several hours, or a tape of meaningless or meaningful sounds put together by a listener on a battered cassette machine. Tapes of eyewitness accounts of a bust or demonstration might be featured, or a recording of a Hopi Indian tribal meeting. Fass often presented live music, and *Radio Unnameable* became known among New York's folk and rock musicians as a place to get together and jam after a night's work. Long before they were popular, such people as Bob Dylan, Richie Havens, José Feliciano, Judy Collins, the Incredible String Band, Arlo Guthrie, and others too numerous to mention, were regulars on Fass's program. (The day after Arlo Guthrie performed "Alice's Restaurant" on *Radio Unnameable*— its first airing—the station received hundreds of calls asking where the record could be purchased. And after Fass had played the tape for a week, other New York stations began receiving as many calls every day. At the time, of course, Guthrie hadn't even signed a recording contract. When he did, the company, with at least partial thanks to WBAI, had a guaranteed smash.) On other nights Fass might just ramble on, free-associating from the innards of his complex mind.

*Radio Unnameable* was then considered to be of negligible value to the station. It was assumed there was little, if any, audience at that time of night, so if Fass was foolish enough to give up his Friday nights to work for nothing, he could do what he pleased. In addition, there was some strong opposition to his kind of program. Some of the old-time members of Pacifica's board, as well as a few of the older staff people, felt that *Radio Unnameable* revolved too closely around Fass's personality, and was too much a reflection of his own viewpoint. They believed the program was too frivolous—too close to entertainment to be in keeping with the dignity of Pacifica Radio.

Those who maintained this attitude, it seems, had made an accurate reading of neither the needs of the audience nor the words of Pacifica's founder, Lewis Hill. Hill, in his essay on the nature of the audience, had written:

*A mutual sense of respect in broadcasting is possible only when the broadcaster does, in fact, honestly participate in his own act— that is, when the thing broadcast actually arises in, or answers in immediate and profound ways, the broadcaster's own sense of value. It is then . . . that we can accept the occasion as real.*

*Radio Unnameable* was one of the few programs on WBAI at the time that I could bear listening to, though I did disagree with many of Fass's opinions. But I was intrigued by the loose format and, particularly, by his very real, personal, and warm manner on the air. He spoke spontaneously and intelligently, often of the way he felt rather than the things he thought, and always as though he were speaking to an individual to whom he felt close and for whom he had respect. This too was consistent with Lewis Hill's philosophy:

*When we think of broadcasting as a series of interested acts, it is plain that the audience, from the broadcaster's standpoint, cannot be conceived other than as a single individual. That individual, being imaginary, is no doubt a compound inner image representing an idealization of the broadcaster himself. So be it. This identification with his audience is also an ethical obligation of the broadcaster. When the address in the studio is to that part of the broadcaster's own identity which he most respects, most of us will be honored to share it.*

Peeking through the curtain, then, will do no good; for the audience is not an aggregate of any description, but a single individual. Listener-sponsored radio rests completely on this premise and its implications as to the proper genesis of programs. It is regarded as a first rule of the project that the persons who formulate and/or perform the materials actually broadcast must be permitted to forage their own resources, express their real interests—and give the real shape of themselves to what is aired, in both substance and manner.

I had never heard radio used in quite this way before. I was accustomed only to the well-modulated voice reading the carefully prepared words, carefully calculated to produce the desired effect— the kind of radio I'd heard all my life.

I sat in the master control room on a number of occasions and watched Fass in the process of creating *Radio Unnameable*. The visual chaos of his one-man operation was only partially reflected over the air. Fass would arrive only moments before the broadcast, loaded down with shopping bags, briefcases, and cartons filled with records and tapes. (In the early days Fass was forced to purchase most of the records used on his program with his own money. The record companies were not yet convinced that WBAI's audience was of sufficient size, or wealth, to warrant our receipt of promotional copies of new records, a standard expectation of radio stations in major cities. Typically, it was not until Arlo Guthrie's "Alice's Restaurant" became an overnight hit after its premiere on WBAI that hundreds of promotional copies from every major record company began pouring in to the station.)

Fass seemed in a perpetual struggle to propel his huge body down the narrow corridor leading to master control. He appeared to have a built-in radar system that was permanently haywire and caused him to collide with every available object with astonishing regularity and a peculiar grace. Even in the control room, the struggle to contain his limbs continued. Once, as his program was about to begin, Fass moved too suddenly in the cramped control room and jarred the gigantic Western Union Naval Observatory clock from its place above the control board (where it had been for nearly ten years), causing it to come crashing down on his head. Bob, who addresses his audience as "The Cabal," simply opened the microphone and, in a voice sounding only slightly more dazed than usual, began, "Good morning, Cabal, the clock just fell on my head. . . ."

*Radio Unnameable* proceeded from moment to moment, with Fass free-associating one record, tape, or thought to another, generally deciding upon the next record or tape only as the previous one ran out. He would open the microphone and talk about something he had seen, heard, or done, then follow it with a string of recorded material relating either directly or abstractly to that theme. Then, just as spontaneously, he would juxtapose other thoughts and begin a new theme, read a piece of poetry, or produce an unprepared collage. Listening at home, one could never know what to expect next. And that, I soon learned, is an important part of the magic of free-form radio.

Often during the course of his program, while tapes or records were on the air, Fass would answer a telephone call off the air. Sometimes the calls were simply record requests, or suggestions, encouraging or discouraging words, or, more than occasionally, a listener in need of someone to talk to. Fass was almost always willing, even eager, to engage in lengthy conversations, and frequently the feelings and ideas generated by these conversations would determine the course of the program. *Radio Unnameable* had a very real and direct relationship to its audience, something lacking in the rest of the media, and even in most of WBAI's programing.

Other producers at WBAI seemed to put hours of painstaking work into the production of a brilliant hour or half-hour program, to have the information only occasionally digested and used, but more frequently, like the proverbial Chinese meal, leaving the listener hungering for something more substantial an hour later.

Fass seemed to be in touch with both the minds and the emotions of his listeners, and, despite the ephemeral nature of the medium, was able to satisfy more than their momentary desires. In spite of the barriers of electronics, or perhaps because of them,

Fass had a more intimate and lasting relationship with his audience than any performer in any medium I had ever encountered.

Throughout the summer and early autumn of 1965, while maintaining a full schedule as staff announcer, I trod lightly about the station, trying to put together the pieces and relate the realities of this peculiar medium to the fantasies I'd had of it for so many years. Technically, this was not easy. The image in my mind was far more complex and glamorous than the real thing, at least as it existed at WBAI. I'd pictured many small studios spread out around the city and a central control room where producers, directors, and engineers scurried about, frantically delegating studios, pushing buttons and shouting directions. The reality, of course, is one man sitting alone in a tiny control room, amidst a mass of tapes and records to be played at approximately predetermined times. Since much of WBAI's programing is recorded in advance, often the man or woman in the control room sits idly for hours, making only the legally required station breaks and notations in the program log. The rest of the station pays little, if any, attention to him. He and the control room are the final step in a long process, and while the end product of that process—the program—is being broadcast from that control room, the rest of the staff is immersed in the process for broadcast later that day or later that month. And so the person in the control room, while seemingly most connected with whatever broadcast is current, is, in many respects, the most isolated from it. He is at the end of the assembly line, putting the product in its package and sending it on its way—a vital role, but one with little connection to the process or the product.

My announcing job was, needless to say, a great leap from the bookkeeper's desk. Still, I felt, there was a good deal more to accomplish. But what? Pacifica's programing, until that time, had generally been divided into three categories: news and public affairs; drama and literature; and music. This presented a problem for me, since I was noticeably ignorant in all three fields. Those who produced programs in these departments, both staff and volunteers, seemed not only knowledgeable, but well ahead of the *status quo* (or at least this was the impression they gave). It was not an unexpected situation, since WBAI had rested its reputation and hopes for financial support upon a small, intellectually, politically, and culturally elite segment of the community. It was not a situation into which I fitted very neatly.

I had narrowly escaped high school only three years earlier, with perhaps the lowest average grade in the institution's shoddy history. Later I had been admitted, through connections, to night sessions at a local junior college, and promptly flunked out. The New York City public-school system was invaluable during my formative years in focusing my attention on what seemed at the time my considerable lack of intelligence, though my teachers and guidance counselors maintained until the end (as their manual must insist they do with all dumb kinds), that I had "potential," and that I was a nice, if somewhat odd, fellow.

Now where, in this WBAI world of sophisticated ideas, could I fit? While I seemed technically capable of program production, the work was less satisfying than I had anticipated. It lacked the direct feeling of communication I had expected of radio. While I greatly admired the skill and talent of many staff producers, I continued to be most fascinated by what Bob Fass was doing, almost unnoticed, during those expendable post-midnight hours.

I continued to spend a good deal of my time at the station hanging around Fass and his hangers-on. The world of *Radio Unnameable* was, to me, a bizarre one, inhabited by musicians, artists, activists, and always a small band of seemingly homeless stragglers, who hoped to find, in their association with Fass, a real-life touch of the magic he projected to them over the air. Some were satisfied, others crushed by the dichotomy between the real man and their own projections of him. In any event, *Radio Unnameable* was more than a radio program: for the true believer it was a way of life, and Fass's devoted listeners followed him, like converts to a new faith, through his own changes in life style and politics.

Fass was not the first at Pacifica to do a late-night, free-form radio program. It had begun years earlier with *NightSounds*, produced first at KPFA by John Leonard, now editor of *The New York Times Book Review*, and continued later at WBAI by Chris Albertson, whose Saturday-midnight program, *Inside*, was the only other late-night offering then on the station's air. But both programs, while theoretically free-form, were more rigidly structured and preproduced than *Radio Unnameable*. Artfully produced collages of music, poetry, satire, and taped montages, both lacked the intimate interaction between broadcaster and audience that so dominated *Radio Unnameable*.

Though I had observed both Fass and Albertson at work, I was hardly prepared for the complexities of doing such a program, and I knew it. Still, when one Friday night Fass fell ill and asked me to do his show, I had little difficulty accepting the offer. I jumped at the chance.

That program must rank among the most forgettable nights in radio history. I can muster only the most meager of recollections myself. For several hours I played only records I'd heard Fass play, and spoke only the words required by FCC law—station breaks on the hour and half hour. But during those hours I painfully experienced the technical and emotional barriers inherent in this new and different use of the medium. They are barriers unseen and unfelt by the audience. Barriers that, should they be detected, lead to the failure of the effort, the breaking of the spell.

How is it possible to communicate experience and feeling isolated from those to be communicated with, shut off in a tiny room absent of living matter, staring at clocks, meters, and machinery, pushing buttons, pulling levers, surrounded by glass and metal? Who are those faceless beings we call "listeners"? What are they doing? Lovers in the act of love-making? The desperate, who in their despair have found a final, comforting, undemanding companion to speak to them? Those who want simply to be "entertained," enlightened, or amused? And where, while dealing with such realities as finding the next record or tape to play, is there time even to consider such things? And what is it, finally, that gives me the right to be on this end of the microphone?

None of these questions had I ever sensed in listening to or even watching Fass at work. All had seemed to fit together perfectly—even the occasional glaring mistakes—and it flowed together humanly and smoothly. Each *Radio Unnameable* was an entire experience, yet all were connected. Fass could transcend the technological barriers of the medium or, at will, put them to work for him. Fass was an artist. I neither knew my ass from my elbow nor was entirely sure there was a distinction.

I made several more attempts at substituting for Fass, each, in its own way, as unspectacular as the first. Still there was an undefinable excitement about this kind of radio, and I was eager for a chance to explore it on a regular basis.

Chris Albertson, producer of *Inside, Radio Unnameable*'s Saturday-night counterpart, had been on the staff for two years, rising in his first nine months from announcer to station manager—a prime Pacifica example of the "Peter Principle." As a producer, he had a unique talent: his programs were a spectacular blend of jazz, preproduced satirical pieces, and artfully edited sound montages. But as station manager, he helped precipitate one of WBAI's most severe internal political crises.

The crisis came to a head in the summer of 1965, when Chris Koch, WBAI's program director, announced that he would leave shortly for a vacation in Paris. Actually he and a number of other American journalists were on their way to North Vietnam—one of the first groups of American newsmen to be admitted to that country since the intensification of the war. He returned later that summer with hours and hours of tape, which he boiled down into a series of hour-long programs, presenting what, at the time, must have seemed a startlingly sympathetic portrait of the North Vietnamese struggle—views which I suspect today would hardly raise an eyebrow among at least a substantial minority of Americans. But back then the war was still a "police action," American soldiers were simply "advisors," and it was not nearly so safe to project views espoused by a growing, but still tiny "peace movement."

A few of Pacifica's board members, as well as the station's single large financial contributor at the time, were angered over Koch's illegal journey, which had been made without the prior knowledge or consent of the cautious Pacifica hierarchy. Upon hearing the programs, they demanded that Koch make certain deletions. It was an astounding violation of the spirit and principle of listener-sponsored radio, a case of self-censorship within an institution whose unbending dedication is to the First Amendment. In addition, the board was attempting to impose its will in a programing matter over that of the staff, which further aggravated years of already tense relations between board and staff. Albertson, fearing for the station's financial life, as well as for his own job, capitulated to the wishes of the board, and insisted that the changes be made. Koch, backed by a good number of the staff, refused.

For weeks tensions ran unbearably high. The business of running a radio station virtually halted and was replaced by midnight meetings, office politics, and back-stabbing gossip. Finally, with the station near paralysis, Albertson called a staff meeting at which we could confront the issue directly and, one way or another, settle it. Hours of mostly bitter and hostile debate led nowhere. Finally (aggravated by the sudden appearance at the meeting of the station's large contributor) Koch rose, followed by nearly half the staff, and left the room. All subsequently resigned, leaving behind them little more than the physical facilities.

Listeners, shocked by management's failure to support the staff, canceled their subscriptions by the thousands, and within a year WBAI had lost nearly a third of its subscribers—the list falling from eleven thousand to seven thousand. It was an astonishing display of the power of our subscribers, and, in its tragic way, a re-affirmation of the validity of a communications medium supported, and therefore controlled, by broad-based community participation.

In the wake of the walkout, programing became pathetically dull, consisting mainly of hour upon hour of BBC-produced transcripts—productions of Greek tragedies, interviews with Cambridge scholars, and organ recitals—with almost no originally produced programing. Even WBAI's highly regarded newscasts became little more than extended versions of what was available at scores of other radio stations—ripped and read directly from the news wire services.

When the dust had settled, Albertson found himself left with barely a skeleton of the former staff, and he had to shoulder as many of the responsibilities of the departed staff as he could handle himself. Recognizing how overburdened he was, I suggested that we split the responsibility of the Saturday-night show, each of us to do it on alternate weeks. And so, on a Saturday night late in the autumn of 1965, one year after I had walked headfirst into the cash-flow sheet, *The Outside* was born. The crisis out of which it came provided the opportunity Fass and those of us just beginning in free-form radio needed to experiment, to create a new audience and base of support for WBAI, and a style of radio that in the next few years would, in one form or another, be imitated by hundreds of other radio stations, commercial and noncommercial, throughout the country.

During its first year and a half, aired first biweekly and later weekly, *The Outside* produced little of any creative consequence. This is about the best that can be said of it. I spoke few words on the air beyond station breaks, and those I did venture sounded so much like Bob Fass that most listeners doubted I even existed. Callers on Saturday night would invariably begin their conversations with, "Hello, Bob?"—a notion which remained a running joke on the program for years to come. *The Outside* consisted almost entirely of recorded music, humor, poetry, and occasional mixes, blended thematically, and, in the beginning at least, artlessly. I based my selection of material primarily on what I'd heard Fass play during the previous week, which I would begin to collect and organize late Friday night. *The Outside*, it is safe to say, was not exactly breaking any new ground in the creative and spontaneous use of the medium. Had there been any stiff competition for that air time, my career in radio would have been a brief one, but so little was happening at WBAI after the staff walkout, management contented itself with the fact that I was working an extra four hours a week without pay and occasionally recruiting a new subscriber or two.

A generally oppressive atmosphere prevailed about the station in those days. Chris Albertson, growing increasingly paranoid since the walkout, kept his ears closely tuned to the station, never hesitating to phone and chastise the person on the air at any time of day or night. The tension was not conducive to creative growth, and this was reflected not only in our programing, but in listener reaction and financial support as well.

Finally, in the spring of 1966, Albertson, under pressure from all quarters, resigned. He was replaced by Frank Millspaugh, a bewildered-looking, bespectacled little man in his late twenties who appeared to be about fifteen years of age. His background consisted of administrative work in civil rights and education. I doubt that he'd ever seen the inside of a radio station until the day he took over the managership. And that, as it turned out, was to his credit.*

Perhaps Frank's most important early contribution to the rebuilding of WBAI was his hands-off policy toward the people on the air. He was willing to support our work, rather than direct it. In retrospect, it seems the minimum to expect from a radio station dedicated to the First Amendment, but a long history of inept management and internal tension and politics had left most of the staff bitter and suspicious of management in any form. Millspaugh —despite the staff's early misgivings and instinctive abhorrence of the mere concept of authority—proved to be a strong and supportive manager, willing to back the staff even when it meant a confrontation with the board, the press, government agencies, or even our own listeners; once the staff recognized that, they seemed finally to be liberated. The station began functioning again, financial support and staff morale began a slow, steady rise, and those of us doing free-form programs finally began to feel free.

Fass chose to spend more and more of his personal and program time becoming involved in the activities of the newly emerging counterculture. Along with Abbie Hoffman, Jerry Rubin, Paul Krassner, and others, he founded the Youth International Party (Yippies), and over *Radio Unnameable*'s air they planned, plotted, and communicated many of the Yippies' activities for the 1968 Democratic Convention.

Earlier, in the spring of 1967, Fass—with four others—organized a "Fly-In" at the International Arrivals Building at Kennedy Airport. It was an event which attracted more than two thousand people and was the first tangible expression of the birth of the

* Millspaugh claims to this day never to have seen the inside of a radio station.

counterculture in the New York area. It was also the forerunner of the now traditional Central Park "Be-In."

During this period *Radio Unnameable* became an all-night electronic gathering place for the young, committed, and confused youth of the middle and late 1960s. So strong was the sense of community developed around *Radio Unnameable* that Fass's call to a "Sweep-In" to clean up on the filthiest and most neglected blocks on New York's Lower East Side, drew thousands of participants complete with buckets, brooms, cleansers, trucks, enthusiasm, hope, and a sense of purpose so powerful that it could not be thwarted even by the New York City Sanitation Department, which, fearing the adverse publicity such an event might draw, had arrived before sunrise and completely cleaned, perhaps for the first time in months, the several blocks intended as the target for the "Sweep-In." Rather than content themselves with the fact that they had succeeded in drawing the city bureaucracy's attention to the long-neglected area and had accomplished their mission without actually having exerted their own physical energy, the group simply picked up their buckets and brooms and moved on to another block. There they spent the day working side by side with incredulous members of the community, and by sunset the area gleamed in the polluted New York evening.

While Fass and *Radio Unnameable* were bringing radio into a closer involvement with the community, Post and *The Outside*—after a year and a half of traveling in the weekend shadows of *Radio Unnameable*—were struggling to achieve an identity of their own. It had been a safe route, but one that I could no longer honestly travel, what with Fass's active entry into the politics of the counterculture. I wanted *The Outside* to become a world of its own, where listeners could seek and find release and relief. I knew also that if it was to gain a genuine identity, it must establish a unique and binding relationship with its audience.

But first the audience had to be defined, and it could be defined in only one way—as a reflection of myself, the "compound inner image representing an idealization of the broadcaster himself," as Lew Hill had put it. Until now I had addressed myself simply to the "BAI audience"—to what I believed to be their interests and concerns—and had felt only the most superficial identification with them.

I had learned early in life to use humor as both a defense and a tool, to understand why I could not adjust, as could so many of my contemporaries, to the demands imposed on me by a set of values that seemed to have no distinct relationship to my own feelings and needs. While I had used comic devices in everyday life since my earliest childhood (and taken much abuse from my parents and teachers for it), I had not been able, until this point, to translate it into a radio program. Recognizing finally the need to move in my own direction—to establish a relationship with the audience as a single individual, rather than a faceless mass—the program began to transform itself into an expression of my own life. My early fears of "overpersonalizing" the program began to subside, partly as a result of Frank Millspaugh's reign, and partly because of the influence of Larry Josephson.

A computer programer and mathematician, Josephson had begun hanging around the station a year earlier, working as a volunteer recording a series of lectures by psychoanalyst Rollo May at the New School for Social Research. Josephson, a rotund, bearded, beady-eyed man whose intense, intimidating gaze could turn the Statue of Liberty into a raving paranoid, had since taken over the morning program, which had been vacated by poet A. B. Spellman after the staff walkout. The hours between seven A.M. and nine A.M. were, like those after midnight, considered largely dispensable. (In fact, after Spellman's departure, it took several months to find someone who would even be willing to get up at that time of the morning to do a program that one imagined would be listened to by only a handful of perverts arriving home after a full evening of perversion.) Josephson began doing the program while still a volunteer, but after a few months, management became aware of his true value and put him on the payroll—at five dollars a program!

His approach to his listeners was staggeringly different from any which had previously been heard over WBAI. He hated them. He hated getting up early in the morning, and whichever side of the bed he exited from, it was invariably the wrong one. His program, in its early days, reflected this wretched, hostile attitude, and consciously as well as unconsciously, satirized the bright-eyed, cheery, smiling-voiced, "Isn't it great just to be alive?" morning radio personality prevalent on just about every other radio station in New York.

Josephson made no attempt to cover up the way he felt; on the contrary, he used it. He was that "compound inner image" that Lew Hill had written about. Josephson felt like shit, and with good reason. He had to rise at six in the morning to face a world which he neither believed in nor fitted into. He assumed that there had to be countless others who rose each day with similar feelings of disgust, and who could only be further disgusted by a cheery voice blaring from their discount-store clock radio, telling them that happiness lay just beyond a dehumanizing, filthy subway ride to a sterile office, to a routine which produced nothing of personal significance.

Though early on Josephson and his *In the Beginning* . . . took a good deal of abuse from old-time WBAI listeners, in the long run it not only prevailed but gained an enormous, fanatically dedicated audience, which he continued to abuse mercilessly. (In fact, Josephson, partially by virtue of the size and fanaticism of his following and partially because of the sheer terror his mere presence brought to any WBAI staff gathering, eventually gained considerable political power within the station.) But no matter how abusive Josephson was, his audience always seemed to come back for more—and they brought their friends, too. It is a phenomenon easily understood: here, for once, was a human being unwilling, or perhaps unable, to disguise his bitterness, and expressing it, more often than not, with devastating wit and intelligence. People could identify with him, for he had come from their world, he knew its inhumanity, and they felt what he felt.

Though I had been doing *The Outside* for almost a year before Josephson started *In the Beginning* . . ., his influence on me was considerable. Several years older, better educated, and more experienced than I, he already recognized that he was, at least partly, the victim of a system, while I, just past my majority, still groped with such basics. With Josephson around, I realized that one could, with a bit of skill and discipline, grope on the air. . . .

It is difficult to say exactly what the purpose of *The Outside* was. Not everyone listened on the same level, or for the same reason. Because of the component nature of such a program, the listening audience must be made up of many small audiences; while there is only one actual transmission, there are many levels of reception.

Many, I'm certain, tuned in simply to hear a half hour or more of uninterrupted music, not even momentarily aware that the sequence and selection of music was leading to or from a specific place. Others awaited nothing but the frequent appearances of Marshall Efron, who provided, I believe, some of the finest moments of spontaneous satire ever broadcast over WBAI's airwaves (see Chapter 3). Or perhaps they looked to the endless, stoned, nonsensical, free-associative, witty and witless Saturday and Sunday nights verbally spent with Paul Krassner.—Nights we went on taking telephone calls so far into the early-morning hours that each of us had to awaken the other from brief naps taken while on the air, listening to our rambling callers. Some tuned in specifically for the telephone conversations between me and the on-the-air callers. Mostly they were battles of wits: some were funny, or moving, or simply bizarre stories; others became *The Outside*'s regulars—they became the show, while I simply faded into the background. (A more detailed account of some of *The Outside*'s "regulars" follows in Chapter 6, for those who can stomach it.)

Several programs were simply outrageous. It would be fruitless to attempt translating them to print, for their outrageousness lay in the fact that they happened, accidently, spontaneously, and live over the radio. More than once I lost my temper over the air—at a guest, a caller, or at my own fumbling inadequacies. At other times I became so hysterical with laughter it became impossible to utter a word. On still other occasions I was so overcome with emotion—sometimes for reasons quite apparent to the listeners, at other times for reasons too private to hint at—that I wept, or simply signed off the air early without an explanation.

Perhaps the historical turning point for *The Outside* came in the spring of 1967. While Fass dwelt upon, and helped create, New York's counterculture, I was spending more and more time recounting anecdotes and experiences of my frequently traumatized adolescence. Some of the audience complained that I was not fulfilling the obligation of a WBAI broadcaster, but others, early on, understood (even before I did) that I was not simply talking about myself, and that the stories were not ends in themselves but vehicles to draw out and share emotions common in one variation or another to many of my contemporaries.

Late one Saturday night in April 1967, several weeks after the successful and highly publicized Kennedy Airport "Fly-In" (which had been initiated and publicized almost exclusively over *Radio Unnameable*), I arrived at the studio carrying a copy of the previous week's *New York Times Magazine*. In it was a piece about the emotional hardships suffered by the obese child and the scars he often carries into later life.

My attitude toward the piece was cynical. It seemed to me that the *Times*, along with other periodicals, kept a backlog of such articles that could be pulled out of the files and published during slow news weeks. My intent, nevertheless, was simply to use some quotes from this piece to help illustrate the universality of my own experience. My interest in reading the piece became secondary when I noted—quite accidentally—that it had been placed, consciously or not, between a full-page advertisement for a chocolate layer cake available at your grocer's freezer and another full-page ad for a device guaranteed to trim away, pound away, or fold away that ugly, unwanted flab. The irony was too obvious to overlook. The corporate mind creates its markets by controlling our attitudes. They convince us that only one kind of beauty is acceptable, and all who do not fit the current image are to live on the fringes of social acceptability. Advertising tells us that food is an acceptable escape, all the while knowing, because they have created the image, that such escape leads only to more intense distress.

How do you relieve the distress? Just a few pages away they supplied that answer. Torture yourself some more, until, if nothing else, your body achieves acceptability. And the cycle goes on, perpetuating the profits of the image makers, and the agonies of those obsessed with the image.

I suggested, with complete facetiousness, that since the "beautiful people" had recently had their moment of meeting at the "Fly-In," it was now appropriate for the rest of us—those hiding behind the shame of our more-than-ample anatomy—to come together in a truly "massive" celebration, a "coming-out," a joyful and shameless acknowledgment of our common bond, a public announcement that we would no longer be driven into darkness by those who set themselves up as the arbiters of beauty. I suggested we call the event a "Fat-In," adopt "Fat Power" as our slogan, and that the event culminate in the burning in effigy of Twiggy, the toothpick-sized English model who, ironically, was that year's idealization of the perfect female sexual object.

I talked a bit more about the indignities suffered by the obese: the humiliation of trying to squeeze yourself into a premolded, plastic subway seat, as your excess spills over into the laps of those seated beside you; the inevitable and painful backlash of turnstiles; the indignity of buying clothing; and, of course, the ever-present awareness of your own physical undesirability. (As a teenager I had to shed sixty pounds before I could view my penis from a standing position without the aid of a mirror!)

Though I had not seriously considered holding such an event, in the days that followed WBAI's switchboard was flooded with calls from people wanting to know when and where this "Fat-In" would take place. By week's end there were so many inquiries that I decided to make the "Fat-In" a reality.

During the weeks leading up to the "Fat-In," *The Outside* suddenly came to life. At last I felt myself becoming that "compound inner image." And the audience reacted appropriately: some of the electronic barriers fell, two-way communication developed, and audience and broadcaster began to merge. Listeners called, on the air and off, to share their own experiences and ideas. Some of their stories were trivial, others funny, and a few were moving. A good number of the listeners perceived—with justification—that the "Fat-In" was a satirical response to "Fly-In." For a while this caused a good deal of tension between Fass and myself. He saw the "Fly-In" as a tentative coming to life of his long expressed vision of a new kind of community, and he seemed to view the "Fat-In" as a mockery of that vision. That had not been its intention. Rather it was an expression of the fact that some of us felt a certain apprehension and skepticism about this emerging "community," fears born, perhaps, of our self-imposed feelings of exclusion from it.

By the time the "Fat-In" became a reality, the small noncommunity emerging with *The Outside* from its previously unidentified existence had created an entire philosophy of "Fat Power." On the air we had, spontaneously and outrageously, written and performed poems and songs celebrating food and flesh, and militantly crying out against the oppression of the obese. We made up and dug out references, some real, some imagined, to the historical contributions of oversized political, cultural, and social leaders.

Finally, on a warm Saturday afternoon early in June 1967, the perpetrators of the "Fat-In" came together, in the flesh, in Central Park's Sheep Meadow, Manhattan's traditional spot for the gathering of the tribes. What looked like five hundred people showed up, not all of whom were noticeably fat, but who were all at least following the rules and "thinking fat." They carried signs addressing the joys of obesity, and wore clothes—again according to the rules—chosen conspicuously to accentuate quantity. Horizontal stripes prevailed as the pattern of the day. Proudly worn tape measures served as belts for some, while others appeared beltless, having bravely molded about them stretch pants and hip-huggers, fashions current to the times.

All brought food—delicious, chemically unwholesome junk food—which all shared and ate with genuine, ostentatiously uninhibited delight. (This was a great breakthrough for the participants, since it is a commonly held notion among overeaters that fattening food, when shoved quickly down the throat in large chunks in the privacy of one's own kitchen, will somehow pass unnoticed out of the body.)

And finally the climax: to the cheers of the assembled mass, the emaciated effigy of Twiggy, symbol of our oppression, was set ablaze.

The event drew an overabundance of coverage from both printed and electronic media, always more eager to cover a light, highly visual feature piece than a hard news story. (The public is more likely to purchase if it is kept smiling.)

The "Fat-In," and the rather bizarre attention it drew to the station, did not pass unnoticed by the few old-time (and old-line) Pacifica board members. One of them (who once, while president of the foundation, had chastised me for an on-the-air reference to *The Outside* as a "show" rather than a "program," saying with a studied pomposity, "We at Pacifica radio do not put on shows. We produce programs") was vacationing in France and sent a clipping from the Paris edition of the *International Herald-Tribune* to the station manager with an outraged note demanding an explanation of this unPacifica-like foolishness.

However, the "Fat-In," and the identification it helped establish between myself and the audience, marked the real beginning of *The Outside*. As Josephson, Fass, and I began to build separate structures based on the same foundation, so WBAI began to re-emerge from the shambles of its internal political struggles. A new, younger audience found the station, felt connected to it, and became committed to it. They supported the station financially, and enabled it finally to rebuild itself in all areas of programing. By the end of the decade this tiny little FM radio station had become one of the dominant creative and innovative forces in the electronic media.

## NOTES FROM THE AUTHOR'S DIARY

SEPTEMBER 3, 1971
LONG VALLEY, NEW JERSEY, "THE GORGE"

*Frank and Robin said they were sick and tired of listening to me whine about how guilty I felt over not working on the book. It's been almost a month since I last even sat down at a typewriter.*

Bob Fass, reading the part of Pooh Bear (type casting) at a 1972 benefit for WBAI at Hunter College Auditorium. (The author played Eeyore.)

At Bob Fass's "Fly-In," International Arrivals Building, Kennedy Airport. It took a while, but the message got through.
(PHOTOS: BERNIE SAMUELS)

The "Sweep-In" ( PHOTO: CHARLES ROTMIL )

Paul Krassner, about to fall asleep, on the air, during *The Outside*. (PHOTO: STEVE POST)

It's a tough night on *The Outside*. (PHOTO: BERNIE SAMUELS)

Fat Power demonstrators gather at my June 4, 1967, Central Park "Fat-In" to watch me burn a life-sized photo of Twiggy.
(PHOTOS: PAUL BUSBY, BOB MILLS)

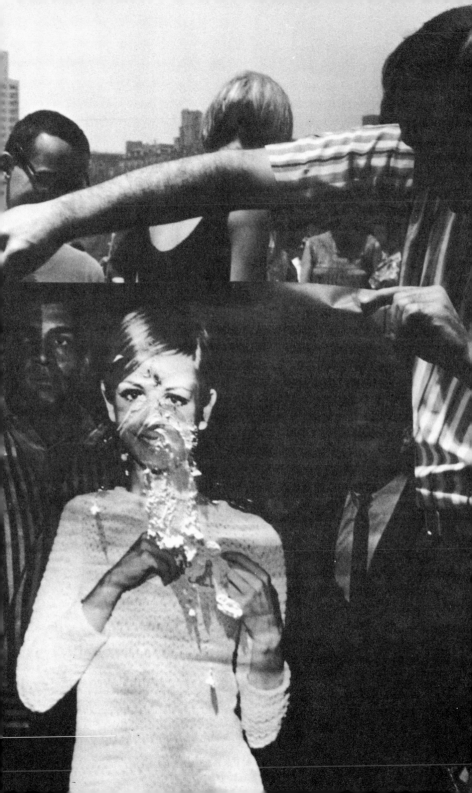

Standing, *left to right:* Dale Minor, Bob Fass, Frank Coffee, Larry Josephson, Eric Saltzman, Baird Searles, Steve Post. *Seated:* Frank Millspaugh. (PHOTO: RICHARD AVEDON)

Larry Josephson  (PHOTO: PETER ZANGER)

Couldn't seem to work myself up to writing about the program, which seems essential before I go on.

   . . . Anyway, they wouldn't even talk to me. Robin put her old manual typewriter on the kitchen table, and she and F. locked themselves in the bedroom and refused to come out until I'd done some work. So I put a piece of paper in and stared at it for a while. It was really a shitty typewriter. Every period made a gigantic hole in the paper, right through the carbon and all. . . .

   Wrote about Fass and the Fly-In, and hooked that up to the Fat-In as the coming of age of the program. Connection works O.K., but doesn't really say anything about the program. At least I've got the sequence of events down now.

   . . . What shit-heads they are—they wouldn't even let me smoke a joint until I'd written at least three pages. It's like being in fucking school again. Now maybe I can go on. I'll try and fill in the spaces next time around. . . .

<br>

---

FEBRUARY 4, 1972

NEW YORK CITY

   It's even harder to get myself to sit down and work here than it is upstate. . . .

   Tried to rework section on the program today. What shit. That crap I wrote at the Gorge last year all had to go—it made the program sound like "programs for young people." (Maybe that's what it was.)

   Nobody will tell me the truth. L. likes everything I write. That's probably why I show it all to her. She thinks I'll throw a shit-fit if she criticizes my work. The fucking editors won't even say anything. There's minor merriment in the office every time I bring in a piece of paper with words on it, like some sort of miracle has occurred. Doesn't take much to keep them happy. Wonder how Viking ever made it publishing indiscriminate crap like this. . . .

   Anyway, I can't seem to express anything coherent about the program. It all sounds like grade-school composition. Did work in some about Josephson and his influence on me (though I was charitable enough to leave out references to the times he tried to, and did, screw me up the ass. But I guess he could claim the same about me). That seems adequate, but I still can't make any sense out of my own program. Try again on the next draft—maybe by then will have gained some perspective, though I doubt it. Seems superfluous to write about what's already been said on the radio. After all, that was the medium, what is the sense of trying to transpose it to print? How do you make sense, or meaning (if there was any), out of the Enema Lady, the stoned talk-fests with Krassner, the recounting of my adolescence, or any of it? It just doesn't seem to work. . . .

    . . . It seems impossible to write about the moment until it's passed, at least. . . .

---

SEPTEMBER 5, 1972
RHINEBECK

Finally got down to writing about the program today. Maybe I've finally got it this time. Possibly all this hurting I'm going through is helping me get it out. This hateful relationship might be having some positive side effects. . . .

Ended the chapter by writing about how Fass believed the Fat-In was a put-down of his life style, how we didn't speak for several months, how it finally forced me to go it on my own, and how the three of us each gaining our own air identity was one of the turning points in the station's history. . . .

It all seems like decent stuff. Just needs finishing touches, which I'll leave for the final draft. . . .

---

MARCH 3, 1973
RHINEBECK

I just can't believe it. How did I ever put such shit on paper? Goddamn those fuckers. Why didn't they say anything—isn't that the editor's job? What are they trying to do, make a complete asshole out of me in print?

Fuck it, fuck it, fuck it. I'll leave the pages blank if I have to. Fuck it. . . .

# 6

# Seven-Second Delay

The use of on-the-air telephone calls between listener and broadcaster should have, and could have, revolutionized the use of radio as a real medium of communication. Instead it has for the most part been turned into another kind of slick format, its prime role being to fill time between commercial messages. Programs are hosted by cello-voiced former disc jockeys whose main motivation is to provide as little spontaneity and genuine controversy as possible and to antagonize as few listeners as possible, so that management will pick up their options. Work in radio is hard to find these days. There have been exceptions, of course, but most of them are unemployed today. Or will be shortly.

On-the-air telephones came late to WBAI—about a year after they were in use at several New York commercial AM stations. Some of the staff was reluctant, having heard what was already being done, and believing that WBAI should be innovating, not imitating. Besides, the best we could do with our limited resources was patch something together out of spare parts. When it was done, it looked as if it had been designed by Rube Goldberg, and the quality of phone calls coming from two blocks away sounded like Edward R. Murrow's broadcasts from London during the blitz.

Our first attempts did not show an excess of imagination. We dubbed the first call-in show *Talk-Back*. Each week it was hosted and produced by a senior producer, and the discussion was confined to a specific topic. With the first half of the program devoted to a discussion between host and guests, listeners were invited to phone in during the second half with questions, or to participate in the discussion. The format was not unique, but the topics and the guests were, in the WBAI tradition, far more provocative than anything generally heard over the airwaves. Also WBAI did not (and does not today) screen telephone calls, a standard practice at all radio stations taking on-air calls. (The stated purpose of this practice is to weed out the "nuts," though too often this excuse is used to weed out anyone whose ideas don't conform to what is considered "safe," and minimize the possibility that the unexpected will happen on the air. All this precaution despite the fact that all radio phone-in shows are broadcast on a delay, giving the broadcaster or producer the option of censoring anything before it actually gets out over the air.) Additionally, callers to *Talk-Back* were not bound by a time limit. "Keeping it moving" has never been a great concern at WBAI— to the station's detriment, some believe.

Perhaps the first meaningfully innovative use of the telephones came during the 1968 Columbia University rebellion, when

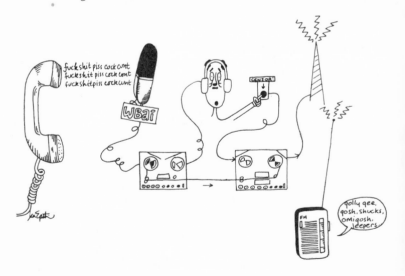

Seven second delay...

WBAI's late-night radio programs (*Radio Unnameable* in particular) became a central communications network for those involved at Columbia, and for anyone else in the city who cared to participate. Live phone calls were broadcast from those occupying the buildings, from the demonstrators and observers outside, and from those calling to attest to police brutality while the blood still gushed from their wounds. One of our own young, long-haired reporters was clubbed down while holding his press card aloft and shouting, "Press!" Some of the police, in an apparent misinterpretation of his cry, proceeded to press their night sticks, with considerable force, upon his head. He had the presence of mind to leave his tape recorder running, and moments later returned to the station, a three-inch gash on his head still unattended to, to broadcast the tape and describe the situation. Then he and Bob Fass took more phone calls from others still at the scene.

For those days, and at other times to come, all participants became reporters, and all reporters became participants. Fass later said that WBAI acted as a giant switchboard, doing nothing more than plugging people, ideas, and events into one another.

(Later, WBAI's news department adopted a similar technique, most notably during the Tombs prison rebellion, when live telephone reports from both hostages and inmates were broadcast. The resulting subpoena from the district attorney's office, and subsequent jailing of Ed Goodman, WBAI's station manager, are covered in greater detail in Chapter 9.)

Toward the end of the 1960s, on-the-air phone calls took a new turn with the introduction of women's consciousness-raising sessions. The innovation of Nanette Rainone, a volunteer producer (who later became a staff member and subsequently, not without a considerable political struggle, WBAI's first feminist program director), CR was the first and at this writing the only program of its kind to be broadcast over the public's airwaves. Following closely in the footsteps of the movement itself, the sessions concentrated on the topic "Marriage and Divorce," and each week focused on a specific idea within that framework.

During the first portion of the program the regular group of women held a typical CR session. The sessions were taped in advance and edited for broadcast, and were followed on the air by about forty-five minutes of phone calls—a continuation of the CR session, with listeners interacting with Rainone and occasionally one or two members of the group.

Reaction to the program within the station was not entirely positive. Rainone took a good deal of abuse from the largely male-dominated staff (myself included), but the tiny woman proved too powerful even for the combined efforts of the male power structure. Eventually her feminist ideas and programing dominated the station—driving out the front door, screaming and kicking, a number of her stanchest male adversaries.

Listener reaction, on the other hand, was decisively affirmative. Women at home during the early afternoon, coping with the thousands of oppressive chores culturally assigned to them, could barely believe their ears. Here, electronically transported into their own homes, was a group of women honestly discussing their lives—lives and problems more likely than not being faced at that very moment by those listening.

CR ran for two years, and along the way probably touched and helped transform more lives than all of WBAI's broadcasting of the previous decades.

During the autumn of 1972, Rainone (who had by this time been program director for almost two years) and I coproduced a series entitled *The Sex Programme,* which used, to a degree, the techniques initiated by CR. Each week the program focused on a specific sexual dysfunction, and incorporated a taped interview with a couple who had been treated in one of New York City's rapidly emerging sexual-therapy clinics, with phone calls from listeners to a participating therapist. *The Sex Programme,* though short-lived (primarily because of the difficulty encountered in finding couples willing to talk openly, over the air, about their sexual problems), was surprisingly well received by both listeners and the local press (which usually conspicuously ignores most of WBAI's programs).

Both CR and *The Sex Programme,* in addition to providing a service unavailable elsewhere in the media, made compelling listening even for those untouched by the specific problems being discussed. The programs provided genuine human drama, undreamed of by even the most imaginative perpetrators of daytime soap opera.

My own experience with on-the-air telephone calls has been somewhat more bizarre. At first *The Outside*'s phone-call segments were fairly structured—callers recounting experiences similar to my own, or relating to a premise I had suggested. One Saturday night, for instance, following my first (and only) viewing of an opera, I conjectured that the reason I had fallen asleep during the second act was my inability to relate anything happening on the stage to anything in my life. I proposed that we compose a contemporary opera, that listeners sing, in any style, accompanied by instruments or *a cappella*, whatever it was they had planned to say. Announcements of demonstrations, discussions of issue, art, movies—anything that normally would be spoken on this night would be sung, as would my response. Thus a two-hour contemporary opera was written, composed, and performed over the telephone by listeners to *The Outside*. It was a unique, if somewhat frivolous, achievement for the electronic media, though the musical world, with the exception of our own music department, took little note of the event. Later that month the music department scheduled the tape for broadcast as a morning concert.

Years later listeners to *The Outside* composed an "Organic Symphony," utilizing only parts of their anatomy as musical instruments, a premise that might have shaken even John Cage a bit.

As *The Outside* evolved, so did the nature of the phone calls. They became less structured, as did the program itself, and slowly a cast of characters began to emerge. Some, like the "Masked Marauder," produced their own radio serials, complete with music and sound effects, and each week, over the telephone, we were treated to his latest adventure.

For almost five years listeners followed, through awkward adolescence, the real-life adventures of another caller, "John." John began his uninhibited, free-associative calls when he was eleven years old, though he had already been a listener for some time. An intellectually precocious child, and a product of a permissive and privileged family, John was very nearly brought up on *The Outside*. He adopted, early on, many of my own speech patterns and associations. As a result, he had an uncanny ability to play my straight man, though half the time he used this ability to reverse our roles.

"Successful" callers tended to develop a following, and, John had his protégés. "Artie" (or "Oddie," as he was immediately dubbed), another preteen, served as a satire of John. "Oddie" was noticeably slow, and had the unique ability to sustain his dullness over long periods of time.

Other regular callers were more specialized, such as the "Hindu-Yiddish Punman," who, as his name implies, authored Hindu-Yiddish puns, which were no doubt highly amusing to an extremely small segment of the audience. Or the "Alpha-bits Kid," who, as the legend developed, ate large quantities of letter-shaped cereal each morning, and at noon spat up peace slogans.

And, each week for more than a year, listeners to *The Outside* followed the true-life adventures of "The Enema Lady"—perhaps the most bizarre series of phone calls ever to be transmitted over the public's airwaves. My first encounter with her came innocently enough late one Sunday night during a random session of on-air phone calls. (No tape of the exchange exists, and the details of it are fuzzy, as are the details of most of the thousands of calls I've answered on the air over the years.) The voice was that of a middle-aged woman, confident, determined, and efficient—like a well-trained, loyal telephone-company supervisor. She had a list of topics to discuss, she said, and proceeded to plunge right into them. All were at least vaguely anally oriented, and they culminated in a cross-referenced list of scatological passages from the Bible. When she finished her list (and not a moment before), she thanked me very much, and hung up. I had barely been allowed a word between topics. My attempts at establishing a dialogue were nearly fruitless, and my instincts to cut off the call before it reached the air were thwarted by my own morbid fascination. She was weird.

As the weeks passed, I continued to receive calls from this strange woman. The scatological references became more specific, and frequent references to what she called the "Penn Enema Club" began creeping into the conversation. My attempts at questioning her were usually ignored, unless they conveniently fitted into her monologue. She was, it soon became clear, an enema fetishist—and proud of it. What was more, she was out to convert the world to her ways. She composed songs and wrote poems proclaiming the pleasures of her anal pastime, described in detail her collection of hundreds of syringes from around the world, and even discussed some of her experiences and techniques. (On several occasions I did cut her off—once when her list of topics included the graphic tale of the "enematizing" of her cat.) Her calls invariably concluded with what she called a "telephone stunt": utilizing the seven-second delay, she would speak one final phrase into the telephone—such as "Give the Holy Ghost an enema"—turn up the volume on her radio and hold the phone up to it, causing the phrase to be repeated over and over, becoming increasingly garbled, and finally disintegrating into a crude form of electronic music.

She assumed her title with pride, and before long began receiving fan mail (as well as, I assume, some hate mail), addressed to "The Enema Lady," in care of a local gay-rights organization in New York City.

Listener response to the Enema Lady was overwhelming and decisive. Many could not believe she was for real, but those who did reacted in no uncertain terms—they either loved her or hated her. Some canceled their subscriptions to the station, and a few even threatened to (and did) complain to the Federal Communications Commission; others sought desperately to find out who she was and make contact with her. Callers claimed they knew her—she was their teacher, school nurse, or neighborhood nut, and I must have been asked ten thousand times if she was a put-on. References to the Enema Lady began to dominate the program, and finally I initiated a Draw the Enema Lady contest and received well over a hundred entries, ranging from crude pencil drawings to elaborate pastel sketches, and even a couple of oils. (Several came in from listeners who apparently couldn't believe their ears, and titled their drawings "The Animal Lady.")

I was more than occasionally criticized for allowing the Enema Lady on the air. Some staff members, and listeners, claimed she was in "bad taste." Though the question must inevitably arise in the course of doing a live, spontaneous radio program, it is impossible to arrive at a suitable definition of the term. When viewed in the light of daily, conventionally sanctioned events, reported without comment or question by the media, the question all but disappears. The Enema Lady, after all, harmed no one, and might even have converted a few to a life of pleasure through "water sports." Whatever their motivations, there was a substantial segment of the audience that could not get enough of her.

In the end (you will pardon the expression), I believe the Enema Lady's frank and good-humored revelations were a positive force in encouraging others to participate later in more serious radio discussions of their life styles. Certainly, it must be said, she set an on-the-air standard for "letting it all hang out."

On the air or off, WBAI gets the strangest telephone calls. We are the first and last resort for everything and everyone. Or so it sometimes seems to those of us who answer the phones. No problem is too large or small, no piece of information too obscure, no opinion too outrageous to bring to the folks at WBAI. So the phones keep ringing.

One night, off the air, Bob Fass spoke to a slightly plastered Christian-American woman, who called to accuse us of being atheistic Communists (or Communistic atheists, I forget which). Fass, acting the patient sage, tried to reason with her, but to no avail. Finally, for the fourth time in the conversation, she reminded him that God put us on earth "to know Him, to Love Him, and to serve Him," to which Fass, after a long sigh, replied, "He sounds like a pretty selfish man." There was a moment of silence at the other end, and then an indignant click.

In November 1971 Fass answered a phone call, on the air this time, from a man who announced that he would shortly be committing suicide. (This kind of call to any other radio station would, most likely, be dismissed prior to air as a "crank," known within the industry as a "dead-baby call.")

Fass, who had no way of knowing if the man was serious or if he was one of the thousands of people who seem to feel a need to use WBAI's air for their own peculiar sort of self-amusement, kept the man on the phone, and the air, for two hours, while the number was traced. Help arrived just as the man passed out from the overdose of pills he had taken prior to calling. Listeners to *Radio Unnameable* that morning could hardly have been bored.

Bad-acid-trip calls have been standard around WBAI for years, and at one point the volume became so heavy that Fass had a volunteer working full-time just to handle them all. At other times we've given out information about gurus, psychiatrists, lawyers, witches, draft counselors, communes, concerts, contests, be-ins, sweep-ins, sit-ins, stall-ins, demonstrations, debates, rides to everywhere, apartments to share, and just about every other real or imagined event, person, or piece of information. A looseleaf book in the master control room contains eight pages of miscellaneous names, addresses, and phone numbers to serve the purpose.

Even with all of that, I wasn't quite prepared for "Debbie," who called off the air one Saturday night late in 1967 and said she was a blind transsexual. She hadn't had surgery yet but for the past year had been taking female hormones and living as a woman, in preparation for the surgery. Recently she had been thrown out of the Lighthouse for the Blind when they found out she was not, as they had believed, a woman. As Debbie explained it to me, "They said that their program was for males or females, and at the present time I fell into neither category. I guess they were frightened I would come to the dances."

It was not your run-of-the-mill sob story. We talked for a while longer, and I gave her the number of Neil Fabricant, then a New York Civil Liberties Union attorney, who I thought would be interested in an unusual case. The problem was simply solved by a phone call or two from Fabricant to the powers-that-be at the New York Lighthouse.

I continued to speak to Debbie from time to time, off the air, and became increasingly fascinated by her and her story. Finally I decided to do a program about her.

At about the same time I received a mysterious phone call one day from an anonymous voice at Young and Rubicam, one of the country's largest advertising agencies. The voice told me the agency was working on a "top-secret, youth-oriented" project, and they were seeking out people they felt had their "finger on the pulse of the youth market." I told them I'd never seen my finger as being particularly on the pulse of any market, but I'd be interested in finding out more about their project. So they invited me to a "top-secret" lunch.

Their aim, of course, was to capitalize on what they saw as the most salable parts of the "youth culture." By mid-1967 free-form radio had gone beyond a few noncommercial and college radio stations and had begun, on the West Coast at least, to show profits in a slightly refined, commercial format. In New York, however, it had gotten off to a weak start, with the financial failure of WOR-FM. The all-new Murray the K was out of work again, and WNEW-FM was making a few tentative steps toward free-form radio. But this change never amounted to much more than a routine format of playing slightly more progressive music than the top-forty AM stations (see Chapter 10). Anyway, Y. and R. wanted a piece of the action.

So, I went to lunch. It was spread in the office of the vice-president in charge of special radio projects, or something like that. The VP had a couple of yes men (and one yes woman) with him, who were referred to around the office as "project staffers." One nervous little man named Ernie actually sat through the entire meeting smiling, shaking or nodding his head (depending upon which way the VP's head was going), and repeating the word "yup." It may have simply been a nervous tic, but it endeared him to the VP.

When it was all over I had been hired as a "consultant" to Super Secret Radio Project X. They were to pay me two hundred and fifty dollars, in return for which I was to produce (or supervise the production of) a pilot tape to be presented to the agency execs, and then to the potential sponsors. The aim, I was told, was a nationally syndicated late-night radio program directed at "the youth." I was assured "complete freedom" both in the production of the pilot and the program, if it got that far. It would be loosely structured, with both taped and recorded segments, and mixed with live talk, satire, and guests. It all sounded too good to be true. And it definitely was.

I set to work on the project, along with Ernie and the other fellow who'd been at the lunch. They were, apparently, mine to do with as I pleased, on loan to me from the VP's stable of yes people. They followed wherever I led, paid for taxis and lunches, opened and closed doors, provided production and recording studios, and pronounced each of my creations a work of genius. I had, apparently, been anointed with sacred oils by the VP.

I had been carrying the idea for "Debbie" in the back of my head for some time, waiting only for the motivation to get off my ass and do it. At the time WBAI had only one small production facility and one studio, equipped with tape machines, microphones, and turntables that had become obsolete ten years earlier. It was a fight then simply to get a tape machine to edit on, and if you succeeded it meant working with five or six others around you editing their tapes, listening to records, or recording programs. It was chaos, and ultimately it took twice as long as it should have to accomplish anything.

I mentioned to my yes men the idea of using "Debbie" as the main part of the pilot tape. They both said yes. Ernie thought the idea had the mark of brilliance. Yup.

A couple of years earlier, during my first days at WBAI, I'd heard an interview with a London secretary from which the interviewer's questions had been cut and the answers spliced together to form a coherent monologue. The program held my attention (not an easy thing to do), despite its excessive length and rather innocuous subject matter. I'd been thinking about trying the same technique for a while, and Debbie seemed the perfect subject.

Thus it was that they escorted me to a plush recording studio located in the garden apartment of a Manhattan townhouse. There we recorded an hour-long interview with Debbie. I spent the next two weeks producing the program—editing it down first into a forty-five-minute monologue, and finally to twenty minutes, with a few classy production techniques I thought might please the slick sensibilities of "the boys upstairs."

I played the tape, which I had been assured by my own "boys" was nothing short of a radio classic, for the VP and his associates. They listened intently, their faces flushing from time to time. When it was over they told me how much they liked it.

"But," the VP added, "I don't know if the boys upstairs will go for it." Ernie, frowning slightly, shook his head from side to side and said, "Yup."

The VP asked if I might not, just for the pilot tape—for the boys upstairs, and the potential sponsors—produce something less provocative. "Something about love, or springtime and flowers." I told him I didn't think so, and left.

A couple of months later "Debbie" was broadcast over WBAI in its original forty-five-minute version, and later was heard over Pacifica's sister stations in Los Angeles, Berkeley, and Houston. Here is a transcript of the program:

INTERVIEWER: . . . *In other words, you feel you were born a female with male organs?*

DEBBIE: *Exactly. That's the way I view the situation. Essentially, what a transsexual is is an individual who has, for some reason— usually many reasons—decided to change sex and is being supported by physicians in this decision. I've always felt that I was a female, period, and just the way I view myself in essence is simply this: I am a female with a physical problem—that is, I look like a male, and this had to be changed if I was going to be happy at all.*

*I maybe had other reasons—I was first raised as a female. I did not know I wasn't until age seven.*

*I was born in Brooklyn and was raised near the Brooklyn water-front—a very poor area—underdeveloped, as they call it these days. As I said, I was raised as a female until just prior to my entrance into elementary school—which made this whole situation come as quite a shock to me [She laughs]. . . .*

*I was with a group of girls and we were playing in the street, and a group of nice, uh, nasty boys came along and decided to rape the whole group and we were taken off to an abandoned house and after they finished raping the first three or four girls, they got to me. And found out that things weren't quite the same [She* laughs]. So they decided to rape me anyway—it didn't seem to make any difference to them [More laughter]. And that's really when I sort of got frantic, because up until that time I hadn't known any difference, I didn't suspect anything. I assumed I was just the same as all other girls and that they were all the same as me. And that was the first time, really, that the actual difference between the sexes was pointed out to me in a physical way.

I was hysterical for quite some time, and I didn't want to accept it at all. I thought I was being terribly misled by somebody—I didn't know who at the time. When I went to my mother, she explained the whole thing away in that it was just a temporary problem that I had, that it would be solved. That I really was a girl in essence, and that the situation would have to be straightened out, and that she would help me do it. But in the meantime we had to fool everybody and continue the big masquerade. If they thought I was a male, we'd just have to pretend that until such time as it could be straightened out. I guess it more or less came as a great relief to my mother that I found out through that means rather than her actually having to tell me when it came time to enter school.

I went through elementary school, and after three years or so—I guess I was just between the third and fourth grade—I was walking around the streets, just messing around, doing nothing in particular except stealing food and things like that, and we found a whole bunch of little capsules in the street. I thought it was a very nice thing. I wanted to find out what was in them. It turned out they were detonation caps, and the one I set off was attached to a few sticks of dynamite and nearly blew the building down that I was in and nearly killed me.

A few months later, when I got out of the hospital, I had excellent eyesight, but my doctors told me that eventually I would go totally blind. Which happened two and a half years later. I went totally blind all at once instead of gradually, as I had expected. I was in elementary school then—nineteen-fifty-six is when it happened—I went to take a book off the shelf and all of a sudden the light went out, permanently.

My first reaction was that somebody else was pulling another joke on me. But after about ten minutes of laughing I started crying [She laughs], and that was that.

Blindness itself is not a problem to me, it never has been. It makes certain things harder to achieve, but it doesn't make anything impossible. There are so many other things in my life that I'd like to have solved. To give you a funny comparison—I know this is an impossible thing—but if somebody said to me you have a choice between which one you'd want: to get your sexual identity determined or your eyesight back. And you had only one choice, and whatever choice you made was a permanent one, I would never even give it a second thought: I'd gladly trade away any opportunity to get my eyesight back, gladly.

I continued in school. By age thirteen I was living in the streets on my own. My family left New York City and told me that if I wanted to finish school I could finish it on my own, which also meant that I would have to survive on my own, which I did. For a year and a half I lived in the streets, stealing food and clothes and all sorts of things, until I met up with some nice guy from my high school who very nicely took care of me for the rest of my high-school career. And I did a lot of traveling with him in Europe. In fact, I was even married to him in Maryland.

He never made any unnecessary demands on me, but I guess he might have been homosexual. But at the time I wasn't in a position to complain, and it was almost a matter of for security you pay a price. And I wasn't in a position to support myself, being a blind person without a high-school education—and I was willing to pay the price. As long as he treated me well that relationship lasted. And he treated me well right up until the time he died, which was just after my high-school graduation.

We had a big argument concerning sexual activity and he vacated the premises and was so angry that he, uh, smashed a car into a telephone pole. Which wasn't very nice. It sort of left me in a heck of a spot, I must admit.

I was attending school as a male, because that's the way I was registered, but outside of school I was living as a female. I went to Syracuse University and at the time was put in a boys' dorm. That situation didn't last long before I suffered a breakdown—I couldn't take the environment. A single girl in a boys' dorm, particularly when the boys don't want to recognize what your sex is, is a difficult situation to handle. After about six months of it I had a breakdown and spent the entire second semester in the hospital.

At the time I had very little understanding of how I could possibly explain something I didn't understand. I came to the conclusion that the people I was dealing with were terribly uncouth and were so immature that even if I were in a position to explain, they would not be in a position to understand. I really resorted to not telling them anything and let them think what they wanted. And obviously they thought some pretty wild things. I knew how I felt, that's all—I didn't know what could or could not be done about it.

Of course, I went to several psychiatrists—mostly who were paid by the state, because as a blind person you are a charge of the state, whether you want to be or not. They determine for you what your future is going to be, whether you want them to or not. They also decide when I should or should not go to a psychiatrist, which ones they should be, and of course it would only be a psychiatrist who was willing to take their fees. Which meant that they weren't very competent. They avoided the issue. They refused to discuss the problem itself, they refused to face what the situation was. They always tried to avoid the issue by saying the problem was something else and this was only symptomatic of something much more severe. A resentment for authority, for instance, or something along those lines, or not having any faith in my elders [She laughs]. These were basically the problems that all this first series of psychiatrists seized upon as the outlet for avoiding what the problem really was.

I think they tried not to deal with me in any way. Invariably they would try and come back to blindness or the problems that were created by blindness. If you go to a sighted psychiatrist and you're a blind person, the problem that's automatically assumed—whether it exists or not—is that you're not adjusted to your blindness. And you have a heck of a time trying to convince them of anything else. Of course, if you try and convince them of anything else, it's because you're avoiding the real problem—which is in essence what they're doing. It's a nice safe problem to say that you haven't adjusted to your blindness, but it's not so easy if you try to face a problem like transsexualism—which doesn't have to be a problem any more. It can be handled, as I found out.

When I was at Syracuse University I had a really good friend from another college. And she was quite understanding at the time. We were always together and everything, and most people always viewed my situation as one of a male, whether I tried to convince them one way or another—they wouldn't accept my interpretation of what I was.

But this particular friend of mine and I really became very good companions and eventually we got married—as a male and female. As to who was who was hard to say [She laughs]. But as I say, Toni and I became very good friends, we seemed to understand each other. She knew about the situation. We didn't know at the time what could and could not be done. She was also a social worker and felt that she might be able to help the situation, and she sort of got involved and so did I, in the sense that we thought very highly of each other—but as friends—and there was nothing ever sexually involved, there couldn't be—I had no interest in a sexual way. From a male point of view, I couldn't possibly ever consider having relations with women—it's something totally beyond my comprehension. It's just not the way I am at all.

After the marriage I tried, maybe if I did my attitude might change. I was willing to try almost anything to see if it would help. And it didn't help, I'll tell you, it certainly didn't help [She laughs]. It certainly made things more complicated, because the harder I would try, the more involved Toni would get—the more she would assume I could do or change things by just deciding I would change things. And then it really got kind of involved, and after three years we separated, and finally last month the annulment was decreed.

But she and I went to doctors to try and find help for me. Eventually I went to my own doctor—my eye doctor—and he referred me to a urologist, and then from him to another doctor, and eventually I got to a Dr. who's a plastic surgeon—a rather notorious one—and he said that he knew of the Benjamin Foundation. So Toni and myself went there to discuss the situation.

I then started dealing with Dr. Wardell Pomeroy—he worked with Kinsey on the Kinsey Report and is now staff psychologist at the Benjamin Foundation. I started dealing with him for a while and he reviewed the background and the evidence, and we went through a lot of physical examinations and came to the conclusion that since surgery was a possibility, I wanted to try it and see what would happen.

About two years after that point, about a year and a half ago, Dr. Benjamin told me what I should do. He told me how I could handle the situation, and I thought it over six more months in therapy, and decided that this would be the right thing for me. I started the hormone treatments, and I've never felt so well in my life—both emotionally and physically. This has been going on now for a year.

I feel that my circumstance is something that would have happened regardless of what my mother did. In many ways I can thank my mother for doing what she did, because it at least helped me over many of the socialization processes that I would normally have to go through. It helped instill the feeling in me of really being a girl at a very early age. That adjustment is very easy for me to make now.

It's simply a matter of a freak of nature, a physical freak of nature. Siamese twins are born, something has to be done, and I happen to be deformed in this way. Now some people consider it quite acceptable, the way I looked as a male, but I personally didn't and this is, I think, the big difference: it was my personally not accepting it for what it was and wanting to modify a situation that I consider was a gross mistake, a freak of nature.

I guess there are a little over five hundred patients now in the world—five hundred who have had the surgery already, but in all of their stories, all of their histories, there are very few points that ever agree. In other words, almost all of these patients have arrived at basically the same point through different means, and it's very hard for the doctors to pinpoint any similarity, outside of the fact that maybe this is the way an individual is born, because the way they have been socialized has obviously been different and many of them have been in as healthy a setting as you could possibly want and still turned out to be transsexuals. The doctors would, on those grounds, say that the indication is that my opinion is right.

But there is really no medical evidence at the present time, or any scientific evidence, to really back it up. Really it's the absence of evidence to the contrary.

When I was in high school most of my friends were girls, with the exception of a few guys I dated. When I was in college, at Syracuse especially, I had very few friends at all. In most cases, I never had to explain. Most of my relationships were so transitional that it wasn't even practical to explain. Those who knew me as a male assumed I was a male, and those who knew me as a female never thought of me any other way. A year ago, when people found out what I was about to do, it cost me all the friends I had. Without exception, all of them. And it's more or less gotten worse as the time's gone by, because the blind community in New York City is a very closed one. It's the type of society in which one blind person cannot do anything without everyone else knowing. And what the majority thinks, that's what goes. I had to not only make a very difficult decision, but I also had to realize that by making such a decision, it was going to cost me all the friends I had. It meant that not only were they not going to be my friends, they were also going to be my enemies. Which is something else entirely. And that they would all work as much against my ever acquiring any friends as anything else. This is definitely what they would do, and it is definitely what they have done. They've succeeded to the extent that they had me removed from attending almost all the agencies for the blind in the city. They complained enough to my office so that I was removed from my job [She laughs] . . . and various other nice friendly gestures.

Well the Lighthouse threw me out—it's that simple. They said that their program was for males or females and at the present time I fall into neither category, so that they couldn't service me [Long laugh] . . . that was their rationale. But the New York Civil Liberties Union has helped me out on that one, and as of last Friday has allowed me to enter their program. Provided that, um, I don't dance with anyone [Laughter]. . . .

I work for New York State, for a parole agency for delinquents, the Community Service Bureau of Metropolitan New York. The agency at first didn't do anything—they didn't know how to handle it. I'd been working for three and a half years and they felt that they would just let it ride. But since they didn't take a firm stand, certain individuals felt that they could criticize, abuse, and call me the usual names that go along with this type of situation. It really deteriorated into a three-ring circus in the office. It really got out of hand. And then with the advent of the telephone calls from the organizing blind people in the city, it made things extremely rough, and they decided that I must be removed from attending the office on a daily basis. Which is what they did. For the last four months, I've been working at home in total exile.

People don't like change—not at all—particularly if it's a change they have to adjust to. And obviously they very much felt they were deceived. They viewed me as a male, they can't see how they can accept me as a female. What they really are saying is that they don't want to, because it means they have to adjust. They might have to treat me a little differently. This fear of change I think is the biggest thing. Rather than try to go with it, they try to fight against it. Particularly since they feel that it's a change of one person—if it were a change of an entire group they might be forced by circumstances to go along with it.

If the attitude of the group changed . . . Blind people, I'm referring to specifically, because the attitude of sighted people has not been as hostile, and that's understandable too, because blind people are very impressionable, and very, very much concerned about what sighted people think of them, and they feel any departure from the status quo among blind people is a blight on the reputation of blind people generally.

Homosexuals generally are the most hostile toward transsexuals. And it's very easily understood, because they personally feel very threatened by what I or any other transsexual would be doing. On the surface, it looks like we're putting them down, as a group. Saying, Well, look, I'm solving my problems, but you can't adjust to yours. This is the way they view our attitude. And that's unfair to us, because my attitude toward homosexuals is simply that, um [Pause], that's their problem [Laugh] . . . I know what mine is and let me handle it my way, and let them handle theirs whatever way they think is best for them, you know?

. . . In terms of what actually is done, first of all there's usually a lot of therapy involved. Usually you go through therapy long before you make such a decision. In fact, none of the doctors involved will accept you as a patient unless you have had therapy. They have to have a full report from the therapist, and if he feels this is the proper decision, he will make a referral to the agency, particularly to the Benjamin Foundation. Then you go to the Foundation, or a competent gynecologist, who then starts administering hormone treatments. You go through many many physical examinations, psychological interviews of all types. Even, in some cases, hypnotherapy—to verify any information that you've given. Once they clear you through all those preliminary hurdles, then they start you on the hormone treatments. They vary with the individual. In my case, I started with pills, on estrogen and progesterone last February. Two months later I started on injections and pills. That goes on for about twenty months. And the changes that take place during this period are a duplication of what an adolescent female goes through, except that it's much much more rapid—twenty months as opposed to six or seven years—so that there is a lot of pain involved. For instance, in the breast development alone, the skin is stretching constantly, the tissue is gathering together underneath the skin, and all this stuff is happening at such a rapid pace that it's a tremendous amount of pain all the time. The thing that you get adjusted to in a very short time is, um, to learn how to sleep on your back [She laughs]. It takes you only about three or four days—all you have to do is roll over once or twice and you feel like you're not going to last very long [She laughs]. . . .

Your weight shifts. A year ago I weighed a hundred fifty-one pounds, I weigh one twenty-four now. My measurements have changed considerably, hair distribution changes, its texture softens —all the usual change that an adolescent female would go through.

All this, of course, is a prelude to the surgery. The surgery itself is, um, uh, how can I explain it? I know a great deal about it from what I've heard from other patients and from the doctors, and it's not very pleasant. It's not a very encouraging thing to learn what you have to go through—the castration and the formation of the vaginal passage, the redistribution of the nerve centers in the urinary tract. All these things are done at once.

There's a very good chance of the patient's not surviving such an operation. The chance is much more substantial than any doctor would like to have—it's about a thirty-to-forty-per-cent chance of death during the surgery, because of the nerve work that has to be done. That's apparently the most complicated and troublesome area. But they do everything now at one time. Several years ago they used to do the surgery in three or four stages and there used to be a lot of plastic work involved. Now they use only skin tissue that's present. For instance, they remove the shaft of the penis itself, but they keep the outside tissue, which is very flexible, and they invert that tissue and use that for the actual walls of the vaginal passage. They use the scrotal tissue for the construction of the lips of the vagina. So there's no waste, absolutely none whatsoever. And it can be done at one time, in one operation that takes anywhere from three to eight hours, depending on the number of doctors involved —and that has varied in various places from one to twelve.

There are, at the present time, about eleven thousand five hundred people who want the surgery in the U.S. and there are only about five outlets that take no more than three patients a month. A lot of patients go outside the U.S., if they can afford eight thousand dollars for South Africa, six thousand for Casablanca. The point is that most of the programs haven't been established. Most of the surgery is being done gratis in the U.S., so the demands for the surgeon's time is so tremendous, and there are so few at the present time involved. For instance, at Johns Hopkins, where they have the largest program, they take two patients a month. They have a very large gender-identification committee—nine doctors involved. They have a transsexual clinic. They have a very elaborate program set up. The surgery costs about two thousand dollars. They have three or four thousand people who want the surgery done. My name is on that list, along with the other three thousand nine hundred ninety-nine or so. They have to take one patient at a time because no individual in this particular case has any greater need for it than the next patient. You can logically find reasons why some patients should be taken before others, but since there are so many, they have had to resort to discrimination.

For instance, Hopkins says that if you haven't lived and worked as a woman for one year prior to the surgery, they won't give you the surgery. This is a means of, shall we say, limiting the number of patients that could even meet their requirements. They next say that if you don't look like Jayne Mansfield or Elizabeth Taylor then they don't take you either [Laughter]. So that eliminates a great number of others. That's how they wind up with their rather small waiting list.

In my case, they turned me down on lots of grounds, the foremost reason the first time being that I couldn't see. Now, for the life of me, I can't understand what that has to do with it. Their explanation was kind of a sad one. Their explanation was that they haven't dealt with blindness before—I'm the only blind transsexual that they know of in the entire world—and that means that makes me a rather unique problem. They don't know what effect blindness might have on how I might adjust, and they don't want to take the chance.

Lots of programs are being set up, but already they're unbelievably expensive. It's the typical mercenary-surgeon attitude—my skill is getting so good that you'll pay, regardless of what I charge. And the funny part about it is there are always patients who will pay. The sad part about it is there aren't many who can. I happen to be one of the latter group.

. . . Well, I must admit, my mother thinks I'm a doll [Laughter] —she wasn't so crazy on the idea a few years ago when she first heard about it, but after she's seen what happened in the last year or two, I guess she figures with the adjustment I've made, it can't be all that bad.

The big thing really is that if you don't have the money, you're forced to wait. If you have the money at your disposal, you can get the things done quickly. Since I don't, I'm sort of put in the situation where I have to wait for those organizations that are willing to take me at a minimum cost, or to have somebody else pay the bill, or to have the surgery gratis—this is what I have to wait for,

I don't have any other choice—I just hope that I'm not already an old maid before I get started [Laughter]. . . .

I see lots of patients at the Foundation all the time. I think one of the spookiest and scariest stories I've ever heard was from a girl who went to Casablanca for her surgery. They don't speak English there. The doctor is French, the nurses are French, and they inform you that if you are going to change your mind about what you're doing, change it before you walk in the door, because once you get in that door, they don't understand English. And they close the door, and they handcuff you and lead you to the room. They lock you in a room that's got bars on it. They don't allow you a second chance, you know, you don't change your mind once you get in the hospital. You look out the window and there's nobody around for miles—it's on the outskirts of Casablanca—and you get the distinct impression that if the surgery isn't much of a success, you could be a very lonely corpse. . . .

The police have a tendency to take advantage of a situation. They feel that who's concerned about transsexuals, homosexuals, and drug addicts, and they throw them all in the same category. They feel that if they want to abuse you they can, it's at their disposal to do so. And if you don't cooperate they have means of taking care of "uncooperative transsexuals," as I was once told by a detective in the Seventy-first Precinct. For instance, driving along the streets, watching your house, or something like that, and when you come out of the house they stop you on the street, talk to you for a few minutes, then arrest you. And you say, "Why are you arresting me?" And one of them says, "Well, you were soliciting a police officer." And I say, "Well, I never did any such thing," and the other policeman says, "I'm a witness." And so you have two policemen, one swearing to the other one's lie. And you have the fact that you're a transsexual, which up until now has not won you much sympathy when it comes to a courtroom. They hold this over your head, so if you don't "put out," then they make things rather difficult. And rather than do that every time I encountered that sort of situation, I moved.

I was at the Benjamin Foundation just before I went to

WBAI, and apparently when I left there, the policemen started following me. I don't know how they went about this, but apparently one of them followed me in the subway. When I got to BAI, I went in and was with you for about an hour and a half. I came back out and was walking down the block and these two people came up out of a car that was pulled over to the curb and told me that they were policemen and wanted to question me and take me to One hundred Centre Street. And I said, "What for?" And they said they wouldn't tell me why, they said they had "questions." I said I wouldn't answer any questions until after I talked to my lawyer. So they said, "How are you going to talk to your lawyer if we don't allow you to make a telephone call?" Which is a very good point [Laughter]. . . . I said, "Well, how do I know that you're police officers?" They said, "Well, you'll have to take our word for it." I said, "What happens if I don't want to?" They said, "Then we can always arrest you for resisting arrest."

I don't really know if they ever have anything legitimate on me or not, so I was sort of put in a position where if I didn't cooperate, I might have gotten in trouble—and I didn't know why. I wasn't quite sure of why I might have, but I never put anything past my friends [Laughter] . . . so I went with them.

It turns out that it certainly was a police car—I know a police car when I see one—or when I hear one, to be more appropriate. So we got down to what was supposed to be One hundred Centre Street, and sounded like it. I'd been in One hundred Centre Street on several occasions—I was married there, and annulled in the same building.

Then what happened was for about an hour and a half they questioned me and questioned me, and I refused to answer questions. They said they might accuse me of being a female impersonator. And I said if they wanted to take the chance on whether or not I was, let them go ahead, but if it turned out I wasn't an impersonator, they'd have a nice suit on their hands. And so they refused to go through any physical searching or examination. So I said, "Why did you really want to question me in the first place?" They said I'd been associating with subversive agencies. Case in point: BAI and the Benjamin Foundation [Heavy laughter]. But after about an hour and a half of really just being annoying, they decided that they couldn't legitimately detain me any longer—and that was that.

After surgery, I expect that I'll do the same thing with my life that every other girl wants to, and that is eventually get married. I will, of course, be working—I'm too independent to stop that— I'll make the money to support myself besides, and pay off all the doctors' bills, which I think I'll still be paying for when I'm retired on Social Security [Laughter]. . . .

But really, I just want to be as happy as . . . I feel that I'm just approaching, right now, the best part of my life, and once the surgery is complete I will have a lot more to offer someone. Because I think that once you're happy as an individual, you can make someone else happy, and not before then. And I feel that up until that time I won't quite be as happy with myself as I should be. Once the surgery is achieved I will have a lot more to offer and I can start leading what I consider the normal life that I have been deprived of up until this time.

. . . I've gotten over everything, I adjust to everything very well. I may joke around about the tragedy, but it's only because, like many other things, you laugh instead of crying, you know? You laugh in many places where if your true feelings were known you'd rather cry. . . . But, um, I don't cry about many things anymore— not in public, anyway [Soft laughter]. . . .

**END OF TAPE**

# 7

# The Anarchists' Circus

WBAI's perpetual and most serious crisis is financial. It was and is the goal of listener-sponsored radio to be financed solely from day-to-day listener subscriptions, and until 1965 they were the mainstay of the station's support. There were outside attempts at fund-raising, both on the air and off. There were small benefits, and well-heeled supporters of the station held fund-raising parties to which they invited their well-heeled friends and guiltified them into forking over semisizable chunks of blood money. On the air there were special days of live programing, interspersed with lots of "pitching." But always it was done "tastefully," Pacifica style—dignified, impersonal, and boring. Not much different from our day-to-day pleas for funds, those standard pieces of informational copy, read in traditional radio rhythm, which sounded about as believable as Nixon's generation-of-peace pitch. The on-the-air events took weeks to prepare, while we lined up guests and programs, and shoveled people in and out of our one tiny studio. For all our trouble, the events rarely raised more than a thousand dollars.

By the spring of 1965, WBAI had reached a critical point both politically and financially. It was before black power and women's consciousness. Between Richard Nixons. When the civil-rights movement was so respectable it didn't want its name hooked up with "peaceniks." It was during the first coming of Love, Peace, and Brotherhood. When pot led to heroin, and Timothy Leary was a college professor, and Tex Antoine was a weatherman.

126

The station itself was about sixty thousand dollars in the red, with some bills going back years. Litigation proceedings were being threatened daily. Chris Albertson was still the station manager. Chris Koch was the program director, and Dale Minor director of drama and literature. Larry Josephson was a volunteer lugging an Ampex 601 by subway to record Rollo May lectures at the New School. Paul Fischer was a cub reporter for the *Hunter College Arrow.* Bob Fass had just become a staff announcer again, having been fired by an earlier station manager. And he'd just started his second round of *Radio Unnameable*—once a week, as a volunteer. *The Outside* wasn't even a gleam in its father's eye. Saturday at midnight *Folio*, the WBAI program guide (when it arrived), offered *The Inside*, with Uncle Chris (Albertson)—certainly the finest late-night radio program of its time. In the morning, at 99.5, you could wake up to the voice of poet A. B. Spellman (the station still gets dozens of mailing-list-crap pieces a day addressed to both Mr. and Miss A. B. Spellman), reading early angry black poetry, giving news headlines off the AP wire, playing jazz or blues, or talking to a guest. It was a weird way to wake up.

At the time I was still the bookkeeper, though the only asset I brought to the job was an ability to bullshit my way around the creditors—a not-to-be-looked-on-lightly ability at WBAI even today. (The same ability, slightly refined, that I've so often applied to my program.) But by spring 1965 I was beginning to run out of heart-rending excuses. One morning I was greeted by a man who, my slightly distorted brain recalls, was wearing gray overalls and carrying a gigantic monkey wrench over one shoulder. He was from Con Edison and was paying a visit, he said, to shut off our electricity. I told him I didn't think that would be a good idea, since we were, after all, a radio station, and would be of little service to the community without electricity.

The simple and pure logic of my argument did not seem to impress him. He insisted that we pay the bill (several hundred dollars), or at least a portion of it, if we wanted the power continued. A day earlier I had written a rubber check to a different corporate supplier in an effort to prevent them from removing another of our vital organs. There was nothing in the bank, and the morning's contributions were barely enough to pay for the kilowatt hours consumed by my adding machine (used mostly for subtracting). Finally, and in desperation, I asked all the staff members around the station to empty their pockets and checkbooks. The man with the monkey wrench thought it highly unorthodox to be partially paid in currency and coin, but after a call to his home office accepted it as a token of our institutional good faith, and left us with our transmitter shaken, but unbowed.

I finally reached my level of incompetence at the task the day the man from the Internal Revenue Service showed up. It may have been his suit that threw me, or maybe his crewcut, or the bulge at his hip. He wanted to know why we hadn't paid our payroll withholding tax for the past three quarters. I told him we didn't have the money. He wanted to know why we didn't have the money we had withheld. I tried to explain to him that we never withheld it because we never had it to withhold.

It was hopeless. He insisted we were breaking the law. I thought about trying to tell him how some laws are morally wrong—but I wasn't sure about that myself at the time. But he was sympathetic —as sympathetic as one would expect a minor government bureaucrat to be. He told me and the station manager we had till the first of July to cough up the thirty thousand dollars or so. After that date the IRS would come in and padlock the equipment and confiscate our assets, such as they were.

It didn't look good. We literally did not have a dime in the bank, with sixty thousand dollars in accounts payable. The summer was coming, the creditors were closing in on us, and income was running no more than a hundred dollars a day. The staffers were at each other's throats. Internal politicking was consuming more of the staff's time and energy than programing. You could tell by just listening to the air.

It was under these circumstances that Chris Albertson called an emergency meeting of the staff, in his office on the first of the three floors then occupied by WBAI on East Thirty-ninth Street. (My first day at work Albertson had told me not to bother filing some papers, because the station would soon be moving. That was in December 1964. WBAI moved to its new headquarters on East

The main entrance to WBAI's elegant new headquarters in a deconsecrated church on East 62nd Street. Our neighbors love us. (The trash cans are filled with yesterday's programs.)
(PHOTO: MONROE LITMAN)

Steve Post, a bit overeager to get on with the show, poses with Pete Seeger at a 1966 WBAI benefit at the Village Theater (later to become the Fillmore East). The benefit also featured Tom Paxton, Judy Collins, Patrick Sky, Leonard Cohen, and the Mitchell Trio.
(PHOTO: BERNIE SAMUELS)

*Left to right:* An anonymous skull, Steve Post, Bob Fass, Frank Mill-spaugh (in their youth), and the back of Dale Minor's head, stepping on each other's lines (and toes) during an early marathon.
(PHOTO: PAUL BUSBY)

"Joselle" the Snakedancer dancing with her snakes for the folks in radioland, May 1968. You may wonder why. So did we.
(PHOTO: PAUL BUSBY)

Sixty-second Street on April 1, 1971. That is the kind of blind institutional optimism, in the face of all contrary knowledge and information, that has kept WBAI barely alive.)

Albertson laid out the problem. Since the staff was small and, at the time, had a superbly functioning grapevine (on which I, as bookkeeper, was an important grape), this news came as a surprise to no one. No doubt WBAI had previously had financial crises during its five years of life as a listener-sponsored station, but none so large and immediately threatening as this. The staff re-examined its past fund-raising efforts, but it quickly became obvious that we had never yet done whatever it was that needed to be done if WBAI was to survive past the month of July 1965.

What had to be done was also obvious—our only recourse was to suspend all programing and turn to the audience, twenty-four hours a day, until either we raised the money necessary to stay alive or the plug was pulled. The staff, however, was not unanimously in favor of this idea. Some could not justify asking the audience to support an institution so racked with internal conflict that it was threatening to decay from within. Others simply believed that our listeners would not support a sinking ship. Still others found singing for their suppers distasteful, and a compromise of the station's and their own personal ideals. This marathon thing more than smacked of Jerry Lewis selling diseased children to the highest bidder of guilt. The staff, then as now, was reluctant, even at death's door, to compromise what they perceived as their "integrity." And, like today's staff, they had difficulty seeing that integrity for what it sometimes is: condescending, elitist, and egotistical. But finally they came through, realizing that if the station was to survive, this marathon, as they unenthusiastically named it, would have to be.

There was little discussion of methods and structures. The goal was set—twenty-five thousand dollars—the absolute minimum necessary for the station's immediate survival. The dominant feeling around the station was gloom. We had no idea how large or tolerant the audience was, and because there were bitter and hostile feelings internally, the staff tended to project those feelings onto the audience.

Finally, it became clear, the marathon had to begin within forty-eight hours. No one on staff had seen a paycheck in months. Con Ed was still threatening to cut our power, and Ma Bell was poised ready to disconnect our cosmic trunk line. Without our friends the public utilities, it was back to batteries, tin cans, and string.

From here on my memory, as it does when looking back on all marathons, gets fuzzy. But fuzzy is the nature of marathons, and the first one set the pace. I can't remember exactly when it started, but it wasn't long after the meeting, during the first days of June 1965. Nor do I remember whose voice began it all—or do I even care, save for the sake of esoteric historical accuracy.

The air around the station was filled with desperation, chaos, and conflict. No one knew or cared what anyone else was doing, saying, or thinking. It was, to use the words of Bob Fass (spoken six years later in regard to a completely unconnected issue), "every fucker for himself." Albertson, who had already noted the rumblings of a staff revolt, engaged himself in a paranoid attempt to keep dissenting staff voices from the microphones, and rarely left the studio during the marathon.

But despite his watchful mouth, others did get on the air. Hundreds of others. Almost anyone who walked in off the street. Artists well-known and not-well-enough-unknown came around during that first marathon. Judy Collins was there, and Pete Seeger, Robert Ryan, Tony Randall, Herbert Biberman, and John Henry Falk. Anyone who walked in with a contribution was brought into the studio, or allowed to wander in. Volunteers, musicians, writers, artists, fading celebrities (Richard Lamparski carried Connie Boswell up the three flights of stairs to the studio), the talented and untalented—all paraded, crawled, pushed, shoved, kicked, scratched, and bit their way in and out of our one tiny studio, some as much to help their own careers as to aid the station.

The three tiny floors were continuously crowded with well-wishing human beings who volunteered their services, brought food, ran errands, and simply hung out. Other kinds of organisms—donated as "barter" items—inhabited the bathrooms and storage closets, and several turtles with HAVE FUN IN MIAMI BEACH painted on their beleaguered shells spent their final days sharing a bathtub with several hundred hours' of used recording tape. Hundreds of mostly useless items lined the hallways. Bodies of all varieties could be found sprawled on the floor, drooped over tape recorders and curled up under the studio table, sleeping, fucking, meditating, and kicking habits—all at once or one at a time. It was an anarchist's circus.

My own role in the first marathon was not central. Aside from counting the money, it was my job to come up to the studio once a day and read aloud from the accounts payable book. It was an inauspicious beginning.

When it was over—about five days later—we had a station piled so high with objects, decaying food, bodies, and general filth, that it never really recovered, and the staff was so tired and bitter that it never fully recovered either. But we had found something that would later play a primary role in the rebuilding of WBAI: we had finally begun to speak to the audience directly, as individuals, rather than simply as voices dropping information on a random mass, employing a technique that was central to the development first of free-form programing and later of programs that dealt with specific human concerns rather than broad intellectual concepts.

And the twenty-five thousand dollars did save the station, or at least keep it alive. Internally though, the downhill slide continued, culminating in Koch's trip to Hanoi, and his subsequent resignation, along with those of seven other staff members. In fact, about the only thing that did survive the debacle of 1965 was the concept of the marathon itself. The following year, under the management of the newly arrived and still thoroughly bewildered Frank Millspaugh, we doubled the marathon goal, began it on May 1, and called it the "Mayday Marathon," after the international distress signal. That marathon began with the simulated sound of a ship's radio broadcasting the mayday distress signal, and throughout the first days staff members repeatedly shouted the mayday call over the air. It seemed amusing and appropriate to us at the time, until it was pointed out that we were breaking the law by broadcasting a false distress signal.

(Some years later a rather obscure rock group put out a song entitled "Oh Dear Miss Morse." Unbeknownst to most disc jockeys and program directors, the rhythmic accompaniment spelled out the word "fuck" in Morse code. When this, along with the fact that Morse code is considered, under FCC regulations, to be the first language of radio, came to their attention, the song was rapidly banned from the air.)

After 1966 it was clear that the Mayday Marathon would become a WBAI institution. Not only was it an effective fund-raising tool, but many listeners claimed it was one of our finest programs. (This may have said more about our regular programing than about the marathon.) Listeners tuning to WBAI for the first time during marathons seemed to become hooked simply by hearing the sound of real human beings trying to communicate a need to other individuals—our frustrations trying to break through the electronic box. I've pictured myself, during countless marathon shifts, curled in a fetal position inside somebody's radio speaker, kicking, crying, screaming phone numbers, begging to be released. The marathon is a cathartic experience for staff and audience, staff members taking the opportunity to express to listeners their feelings about them, sometimes in the most basic ways. It is clearly, for most of us, a love/hate relationship, the balance changing drastically with each shift, fluctuating with the success or failure of each appeal. Egos and emotions run rampant over the airwaves. Some shifts sound like the tantrums of petulant children, others like the ravings of the stereotypical Jewish mother.

The newer staff member having his or her first "marathon experience" will sound immensely reasonable, discoursing on the First Amendment and WBAI's dedication to it; discussing, with illustrations, the changes in programing, the growth of the station; detailing our service to the community, our impact on the rest of the media, our importance to and influence over humankind and the universe. Immensely reasonable. But, no doubt, completely ineffective. And extremely frustrating. And this frustration, often translated into resentment toward the unresponsive listener, can be channeled into effective, creative, and very human marathon radio. It is a liberating feeling for the person behind the microphone. Your guard is down, you are angry, or hurt—this is what the audience responds to, these are feelings they know well.

Poverty also creates a feeling of perpetual crisis and constant sacrifice within the station, a state in which the members of the staff, despite their occasional protests, seem to function most effectively. They will curse and bemoan poverty, but they will, in the end, work together to see that the station remains alive. For whatever political and artistic differences staff members have, and they are great, it is this common belief in pure listener-sponsorship—the direct relationship between audience and broadcaster—that binds them together.

Poverty at WBAI is not a temporary condition; it is a way of life. The staff at any Pacifica station believes that poverty is a virtue. (They have no such feelings about other Biblical virtues.) They believe, with good reason, that what they are doing will be mysteriously corrupted the day there is more cash on hand than outstanding debts. In point of fact this happened only once during my stint as bookkeeper, a hideous condition that lasted less than an hour, as I recall. It has since occurred only in the days immediately following marathons.

The perpetually poor state of WBAI's finances is, I believe, institutionally healthy. It indicates a willingness to gamble, to go ahead with bold programming ideas despite the lack of resources to pay for them, and a continuing faith in the potential support of the listeners.

---

## THE 1972 MARATHON AWARDS! *by Ira Epstein*

JUNE 2, 1972

*A resonant voice echoes through the sound system in Consolidated Laundry's Grand Ballroom as the staff, management and friends of WBAI await the start of the 1972 Marathon Awards Presentation:*

Voice: *Out of the red ledger book comes a face so shocking with a story so revealing that people shudder and turn their eyes. With the dimes thrown at his feet, the comptroller came upon us, dressed in transfixing invoice paper, peering into a black cloud on a dark night. And with him, that man brought —pitching, shlockstorms, come-ons and "meet Ioar"s, a dirty brown-nose, and a television. Dung donged to the damage of*

*his dawn, and din and dial to the dicker of his drawl: that's why*
*I say:*
GOOD EVENING AND WELCOME TO THE 1972
MARATHON AWARDS!!! *Ladies and gentlemen, Mr. Ed*
*Goodman!*
The crowd is on its feet as Goodman lopes to the lectern, dressed
in a dungaree tuxedo jacket, bow tie, short pants and suspenders.
The applause is thunderous.

Ed: *Thank you. Thank you very much. Gary, you may be*
*seated—we're not making any staff appointments for at least a*
*couple of months. Thank you, thank you, everyone. I had a*
*few words to say about listener-sponsored radio, Pacifica's Great*
*Experiment, oppression and all that stuff (I've been in jail, you*
*know) but Marnie said I should just come out here in my*
*shorts, get a few laughs, and go back to the office to finish*
*pasting those stamps on the renewal letters. Oh, and Carolyn,*
*dear—where are you, Carolyn? I see you, hiding under that*
*table over there. You needn't wait up for me tonight—got to*
*sweep up studio C for tomorrow's Great Pacifica Mindless*
*Lemonade Slurp and Self-Indulgence. Leave the night light on*
*for me and set the electric blanket on "warm as toast"; O.K.?*
*And now, here to present the first marathon awards and intro-*
*duce the award-winning performers is my right-hand man—*
*NANETTE RAINONE!*
The applause is tumultuous as Ed Goodman's male nurse helps
him into his wheelchair and rolls him offstage. Nanette marches
to the lectern, climbs atop three telephone books and speaks.

Nanette: *Thank you so much, Ed; so very good of you to come*
*—and in your condition. Women and men, you're all well aware*
*of the sacrifices Ed Goodman has made for the cause of free*
*speech. We all remember that grueling night he spent in the*
*Executive Suite of the Men's House of Detention, robbed of*
*his vitamin pills, bow tie and electric blanket. Well, tonight*
*we honor those sacrifices by presenting these statuettes, the*
*"Edsies," to individuals who have made outstanding contribu-*
*tions to the success of this year's $215,000 marathon.*

*Just a word about the balloting.*

*The nominees for each award were submitted in writing*
*by marathon listeners. These nominees were then discussed*
*rationally and intelligently at a recent WBAI staff meeting. A*

secret ballot was held and tabulated by Osorio Potter with the aid of a non-military-industrial-complex adding machine. The results were submitted to me by registered mail in sealed envelopes made from recycled paper. And finally, those letters remained sealed in my office desk, in the interest of objectivity, as I sat down and chose the winners.

And now, to present the first inconsequential awards, is the Ghost of Marathons Past, a man who has built his career around marathons and boredom, Mr. Steve "is he still alive?" Post!

Post walks to the platform and stands, dwarfed, beside Nanette. His hands are shaking, his voice quivers and his eyes dart back and forth.

Nanette: Steve, are you nervous?

Steve: No, no, Nanette.

The crowd is tossing paper napkins at Steve, some carefully wrapped around huge rocks and broken bottles.

Steve: Can we please have some quiet out there? Look, I'm up here trying to make a presentation, you know. If there aren't at least ten of you out there willing to shut up, you know, then maybe I shouldn't be up here. You know, there's nothing that says I have to be up here, making these awards. One day you may just wake up and find me gone, you know. O.K.—I take this as a personal insult, you know. I'm just going to be quiet until you get off your asses and respond. (Several moments of silence.) Look, I'm getting paid for this. I don't care if you listen or not, you know. I'll just make my presentation and get the hell out. This first award is for costume design. And the nominees are:

Tony Elitcher—for the bright yellow "69" sweatshirt he wears at every WBAI benefit concert so he can be inconspicuous as he moves microphones, adjusts amplifiers, combs performers' hair and marches up and down the stage in the middle of the performance.

Barbara Oka—for the sleek, svelte, "Millie the Model" one-piece, high-fashion, long-length, shimmering gowns she wears to Free Music Stores, Square Dances and Hay Rides.

And Charles Pitts—for daring to be different.

And the winner is? (Nanette whispers in Post's ear.) Charles Pitts.

*Of course, he had to win. Tony doesn't work here any more, and Barbara's in Prospect Park, taking sound levels on pigeons. Well, come up-Chuck, and claim your Edsie.*

Pitts shuffles to the lectern, dressed in his award-winning outfit: black leather jacket with shiny gold epaulettes, white dungaree pants with a cat-o'-nine-tails and blackjack hanging from the pockets, white tennis sneakers, a sword hanging from a chain around his neck, a whip in one hand, a chair in the other. He speaks.

Charles Pitts: *You can take your Edsie and shove it.*

Steve Post: *Thank you, Charles. Short, but sweet. For our next award, "Most Adequate Performance by a New Staff Member" we have just two nominees:*

*David Selvin—for his realistic "I am under arrest" routine, a sure-fire sympathy-getter around here. And Jim Irsay— for his winning role as Steve Post's protégé.*

*And the winner is, Nanette? JIM IRSAY!*

Jim Irsay stalks to the lectern, cracks his knuckles and speaks.

Jim Irsay: *Well, I'd like to thank all the little people out there who helped me win this award, especially you! You! That fellow in the plaid shirt! You know, this is a really good award, one of my favorite awards of all time, sculpted by a man you probably never heard of, but who was very big in his day. This award is a classic award, an award I hoped to get for a long time, and an award I'll look at for a long time. So without further ado, I'll take this award from you right now. Here goes. I'm about to take the award. I'm cueing my hand onto the award and lifting it off the lectern. Now. Here goes. I'm taking the award. Right now.*

*Steve, how do I get out of this?*

Steve Post: *Not bad, Jim, but you're still popping your p's.*

Nanette: *Thank you, Steve, Jim. Now, before I present the award for "The Most Amusing Marathon Shift" let's listen to some of the nominated shifts. First, we'll hear Frank Coffee and Liza the-woman-with-the-sour-cream-voice Cowan doing a scene from their lilting Babes on Broadway shift. (Music concludes.)*

Frank: *Yes, wasn't that lovely. That was Vera Hruba Ralston making her combined singing debut and farewell performance in the 1913 Noel Coward Revue, Grandpa Was a Lady.*

Liza: *That was lovely, Frank. My mother took me to see that one.*

Frank: *Yes, Liza. The phones are lighting like the lights on Broadway lit during the 1887 Kurt Weil romp Broadway Lights Light Nightly, so I'll just segue into one of my hilarious show-biz anecdotes.*

Liza: *My mother always wanted me to go into the show biz.*

Frank: *Yes, Liza. They tell a story about Ethel Merman—*

Liza: *She's one of my favorites, Frank.*

Frank: *Yes, Liza. They tell a story about Ethel Merman—she was touring in an 1865 remake of that Eastern European classic of light cavalry, Annie, Get Your Hun—*

Liza: *That was a good one, Frank. My mother took me to see that one.*

Frank: *Yes, Liza. Well, she was backstage at the Attila Theater, getting into her corset, when her dresser came into the room with a swatch of material and asked her if she had a match. Without batting an eye, the grand old lady of American theater replied, "Yeah, your ass and my face."*

Liza: *That was a good story, Frank.*

Frank: *Wait a minute; I'm not sure I told that right.*

Liza: *My brother writes for the Village Voice.*

Frank: *Yes, Liza. But before we do that, let's hear a cut from a rare LP I've got of the great Jan Clayton singing "What Will I Do When My Man Walks Out on Me While I'm in the Family Way" from the 1734 musical smash Jub-Jub. Action!*

Nanette: *Cut! Cut!!! My, my, but that was amusing. Now let's listen to a short scene from "The Merry Punsters," starring Larry Josephson, David Rapkin and Paul Fischer.*

Larry: *Welcome back to the Marathon, brought to you by Pacifica Foundations, the makers of West Coast Bras and Garters.*

David: *Bra-ra-ra boom-tea-ay. Garter my Heart, bring back that melody.*

Paul: *One sells locks, the other locks cells.*

Larry: *I'd loaf some bread right now.*

David: *Born to be bread.*

Paul: *One jails hacks, the other hails jacks.*

Nanette: *Cut! Cut!!! Cut!!!! The winners are Larry, David and Paul—The Merry Punsters.*

*They walk to the lectern. Larry is eating an apple as David watches the lights reflect off the Edsie and Paul records the presentation on his Sony.*

Nanette: *Boys, what's the secret of your miraculous punning ability?* (Long silence, until finally:)

Paul: Good evening. As a reporter, I feel it my duty to report that Larry, David and myself, Paul Fischer, spent days preparing lists of complicated puns and double entendres. By merely steering the conversation into one of our "pun alleys," we were assured of at least thirty minutes of sustained word-play. Thus, although we earned approximately three dollars in pledges during our combined fifty-hour shift, we succeeded in bolstering our image, while boring our audience stiff.

Larry: Nanette, he doesn't know what he's talking about. You'd be surprised what a few well-placed rubber bullets can do to a man's brain.

David: Look at the pretty colors.

Nanette: Thank you, boys, you may be seated. And now, we present the "I'm Not Going to Beg You" and "I Shouldn't Even Be Doing This" Awards to Deloris Costello and Charles Pitts.

Deloris: Yeah, thanks, Nanette. I shouldn't even be accepting this award, 'cause it's just not my way of doing things, but I'll do it anyway. Hey—listen. If anyone knows where I can unload three hundred thousand "Free Angela" buttons, will you call me at the station during business hours? I would say thank you, but it's you who's to benefit, not me. Tu-ta-na-na.

Charles Pitts: You can take your Edsie and shove it.

Nanette: Well, all right. This years "Marathon Rip-Off" Award goes to volunteer H. E. Weatherman, for ripping off three hundred records, five microphones, Milton Hoffman's desk, the switchboard, and the balcony of the church. We'd give you an Edsie, H.E., but it's been ripped off.

And now, before we present the fabulous Mr. or Ms. Marathon Award, a brief list of minor winners:

The "Floor Matt" Award to Matt Alperin for being the most ignored person in this year's marathon. Conservative figures estimate he was ignored not less than 1298 times during the three week marathon. For good reason.

The "Star Is Born" Award to Dorothy Barter for reading off lists of broken phonograph records and used flavor straws with such sensual attachment that listeners either reached spontaneous orgasm or fell asleep.

The "Never Say Die" Award to Liz, who threatened to quit three hundred times during a single shift.

The "I Am Going to Be Sick" Award to Marc Spector, for vowing to be violently ill on the console during each of his shifts.

The "I Am Sick" Award to Dan Kavanaugh.

And finally, "Nice Guy" Awards to individuals who maintained a calm, relaxed, friendly demeanor no matter what the emotional or mental strains of the hectic marathon: Caryl Ratner and Ken Volunteer. These people are either extremely nice or lobotomized.

And now, for the award we've all been waiting for—the ultimate Ego Trip for BAI staffers: THE MR. OR MS. 1972 MARATHON AWARD.

And here to introduce the nominees is one of the nominees himself, Gary Fried!

Gary Fried: Why, thank you, Nanette. You look lovely this evening. Fellow workers at WBAI, there are just two nominees for the Mr. or Ms. 1972 Marathon Award—me, Gary Fried, and somebody else. I've commissioned this short film demonstrating my dedication to the goals of WBAI, and I'd like to show it right now. Would you roll it, Charlie?

That's me sitting down in tally, adding new pledges with my left hand, changing the tally with my right hand and answering the switchboard with both feet. At this point I had been up for 72 hours without even a short nap or a bite to eat. I've just computed the new tally, and am now running upstairs to man master control while Steve Post goes to the bathroom. That's a quick cut back down to the tally book, where you'll notice I raised three hundred dollars in just ten minutes on the air and not totally from relatives, either. And now a fade to the Free Music Store, where I'm soliciting contributions and seating people (after having been awake for 96 straight hours, without a bite to eat)—and wait, wait, do you see that man in the corner handing me that hundred-dollar bill? I'm running down to tally now, pushing Ira Forleiter out of the way and taking charge of operations again. There, a quick splice, some twelve days later, I'm still at it, never slowing down. I'm carting Coca-Cola empties back to Grand Union and here I am returning with the $3.12 to plunge into a new tally. After 1215 sleepless and eatless hours I'm making coffee for volunteers, sweeping the basement floor and feeding the station cat with what was supposed to be my supper: a crust of bread. And here the film fades out in a long pan shot of me running, writing, calculating, not eating, not sleeping, not caring about myself, dedicated only to the task at hand—running the marathon singlehandedly.

Well, there you have it, folks. Just my average dedication to the wonderful goals of listener-sponsored WBAI. That's me, Gary Fried, never tiring, never stopping to think of myself, Gary Fried, never complaining, never quitting. And now, Nanette, if you'll just whisper the name of the winner of the Mr. or Ms. Marathon Award into my ear, here—

Robbie Barish?

Like shit! I QUIT!

Ira Epstein
130 E. 39th St.
Brooklyn, N.Y.
No ticking packages will be accepted.

# 8

## Pitts and the Pendulum

The process of becoming a staff member at WBAI is, as has been noted earlier, rather bizarre and almost undefinable. It varies, in degrees, with each individual, but always remains something other than a standard bureaucratic procedure. Paid jobs at the station are scarce, the demand high, and each position, no matter how seemingly insignificant or poorly paid, is fiercely competed for. There is that category of BAI employees I have loosely labeled "Pacifica Orphans"—those who arrive at the front door apparently possessing little other than enthusiasm and need. They may or may not have worked in radio before, but they have invariably been attracted to the station by listening to it. They rarely bring with them coherent programing ideas or concepts. This is not to say they lack talent or ability, just that they simply have not disciplined or defined it. In such cases it may turn out to be the perfect place for them, since WBAI rarely operates by predetermined standards, and it is possible for a young wanderer to roam about the station, as in my own case, trying a bit of this and a bit of that until a place, if any, is found. Often a place is found among the station's people before one is found within the institution's needs. That is to say, it is simpler to find friends than jobs at WBAI.

Though many job applications arrive daily, few, other than those with specific administrative or clerical skills, are ever considered. One reason for ignoring so many applications is that previous job experience is more often a liability than an asset, since

WBAI's unique needs frequently require the unlearning of more traditional radio or journalistic experience and habit. Equally important, though, is that the members of the staff are collectively overprotective of "their" radio station and its people. A volunteer, once accepted personally by the staff or a segment of it, is likely to be considered for any opening on the paid staff, with or without the appropriate qualifications. These friendships or alliances, which have always existed around the station, can be based on any number or combination of things: emotional or sexual attachments, or politics, philosophy, and the sharing of specific skills or interests. Once such bonds have been made they are difficult to break, since the institution fulfills, in varying degrees, both emotional and intellectual needs. Often the two become confused, and the victims are both the ideals of listener-sponsored radio and its programing. In the name of compassion, too many people have been kept on staff too long, frustrating attempts at improving programing or services in their particular area. My own experience as WBAI's bookkeeper is a tiny example of this syndrome. WBAI is as much a halfway house for the alienated as it is a radio station.

Theoretically, there is a standard bureaucratic hiring procedure. The station manager is hired by the board of directors, the manager hires a program director, the program director, with management's approval, hires department heads, who in turn hire those to work within their departments. But management's approval is merely a formality: in theory, both the program director and the department heads are free to hire whomever they choose. On paper, it sounds like a smooth and efficient operation, but only on paper, for the staff has an unwritten, organic, almost barbaric rite of scrutinization it must apply before a prospective staff member is welcomed within the tribe.

First, one must be sufficiently humbled and awed by his or her mere association with the institution. One must be willing to work almost night and day for a considerable length of time, at the most menial of tasks, for no pay. Then there is the unspoken "instability test." If the newcomer seems adequately alienated, and emotionally in need of closer ties with the station and its people, his or her chance of acceptance by the current staff is increased. Without this informal, often humiliating process it is difficult to get hired, yet if someone should happen to get hired without it, he finds it almost impossible to function within the institution.

Many of WBAI's staff have been excluded from or have left society's more established institutions. Some have had what might be considered "successful" careers prior to their association with the station. Larry Josephson, WBAI's morning man from 1965 to 1972, was a mathematician and computer programer, earning more than twice his WBAI salary. Charles Pitts, who produces programs for the gay community, worked for many years in commercial radio outside New York City. Both began at WBAI simply by "hanging around," by being present, and by being willing to assume almost any task not taken up by an established staff member. Remember, for the sake of this "Pacifica Primer," that determination and insistence pay off.

Let us take a close look at the history of Charles Pitts at WBAI, for it illustrates the point best, or at least at its most extreme. Sometime during 1967 I arrived for an evening announcing shift and found, to my irritation—though not my surprise—a thin, bald-headed young stranger standing silently in one corner of the control room, staring threateningly about. (Such an occurrence is far from unusual at WBAI. New faces seem to appear from the woodwork daily. Returning to the station after even a brief hiatus, one can find literally dozens of them, each scrambling about and performing tasks as if he'd been preparing for them all his life). I assumed the stranger was an acquaintance of the announcer on the previous shift, and that when he left he would take his friend with him.

It soon became clear that this was not the case. The stranger, like hundreds before and since, had simply wandered in off the street, either having listened to or heard of WBAI, and had found his way to the control room. The staff members present at the time, who were all wrapped up in their work and anyway dubious of the slightest contact with strangers at even the most relaxed of times, ignored him. (Of course he might have been carrying a bomb, or intended some other kind of destructive behavior— an often-articulated rationale for the staff's suspicion of outsiders, but that seems only to be a verbal paranoia, since these same staff people are generally reluctant even to question an unrecognized body.)

For a while I sat in the control room going about my work. Like everyone else, I acknowledged not even slightly the stranger's presence, hoping he would pick up my not-so-subtle message to get lost, quietly leave, and wander off to another part of the station to intimidate someone else. But his persistence prevailed, as it has many times since, and before long he told me his name was Charles Pitts, that he had worked at a number of commercial radio stations in New York State, detested the limitations of commercial radio, and hoped to do volunteer work at WBAI, where he supposed things were different.

Neither Pitts nor his story was unique. His manner seemed a bit odd, but oddballism is, after all, the standard at WBAI, if not a prerequisite.

Before very long, Pitts was doing volunteer announcing and engineering. Shortly thereafter, having secured the anointment of the tribe, he joined the staff as an announcer.

Pitts at first appeared to have no special contribution to make to the station. He proved to be a competent announcer, though his defiance of the limitations of commercial radio turned out to extend even to the few guidelines then in force for WBAI's announcers. Charles chose to interpret the term "free radio" in its broadest sense: he felt free to do whatever his roving psyche dictated during his announcing shifts, which included taking on-the-air phone calls when he should simply have been introducing the next scheduled, taped program on at least one occasion.

Shortly after Pitts became a staff member, I, as chief announcer, issued a memo suggesting that the other announcers provide more variety in the selection of music for miscellanies (a miscellany being an open slot in the schedule which the announcer traditionally fills with his or her own choice of program material, usually music). Our announcing staff at the time was young and relatively inexperienced, and it seemed to me and to a number of other senior staff people that we were programing an overabundance of popular music in these slots. The memo, without setting specific guidelines, reiterated the traditional ones, which called for greater variety and asked that announcers create smooth transition between programs, without commenting on the content of the programs.

The day after the memo was issued Charles resigned. He said he couldn't abide by such restrictions on his freedom of expression. After all, was this not "free radio"? But he remained around the station as a volunteer, doing mostly odd technical and engineering jobs. After several months and hundreds of hours of discussions—which might more aptly be termed "negotiations"—I rehired Pitts, not to the universal delight of management and the senior staff.

Pitts toiled on as an announcer, teetering always on the brink of those guidelines, and his job, making no major contribution to programing.

In the spring of 1969 Baird Searles, our drama and literature director, who had never kept his sexual preference a secret within the station, initiated a series of programs entitled *The New Symposium*. Described as a program "of, by, and for the homosexual community," it was the first of its kind on WBAI (though there had certainly been many hours of astute discussion of the subject) and, I more than suspect, the first of its kind anywhere in the electronic media. The program was hard won even at WBAI, against the initial objections of the program director and other senior staff people. Pitts involved himself in the program and before long became its coproducer.

In June 1969 I went to Los Angeles to work for a while at the Pacifica station there, temporarily vacating my Saturday- and Sunday-night programs. At the same time, Bob Fass took an extended vacation, which opened up the Monday-through-Friday night slots as well. As had become the rule, our programs were rotated among just about everyone, staff and volunteers, who expressed a desire to do them, and spoke enough English to identify the station. Pitts qualified.

He began his first program by saying simply that it had become important to "live" free-form radio for the personality to relate honestly to the audience, and for the audience to relate similarly to the person on the air; therefore he would feel more comfortable if he told the audience at the outset that he was a homosexual. Then he played some music and later opened the on-air telephones.

The audience response, as might have been expected, varied widely. There were those who were offended and others who were simply intrigued. Most significant were the calls from those in the audience who were gay. For the first time there was someone on the public airwaves with whom they could identify. They called on and off the air throughout the night, trading histories with Pitts in something of a mass electronic catharsis. Some emerged from the closet that night for the first time, before even having made the admission to those closest to them. (I should point out that all this took place before the gay community became visibly organized.)

A couple of weeks later Charles filled one of Fass's weeknight slots and again focused the program on homosexuality. I was still out of town and so heard neither of the programs, but I strongly suspect that by the second one Pitts had gained a degree of confidence which made his discussions less theoretical and general and more personal and graphic. I also suspect that though the telephone calls were broadcast on delay, Charles' reluctance to be bound by even the most minimal restrictions kept his finger far away from the "censor" switch (which is today attached to what is commonly referred to around the station as the "fuck speaker" and which enables the engineer on duty to cut anything off the air and still be able to hear it in the control room).

In any event, I phoned the station some days later and spoke with one of the senior staff people who, echoing the sentiments of others, seemed less than enthusiastic about the Pitts programs. Their main articulated concern was the FCC, and the notion that the commissioners might not look favorably upon such programs at license-renewal time. Pitts maintains that these staff people never voiced their concerns to him.

When I returned to the station a couple of weeks later, I found Pitts submerged in still hotter water. "Rock" miscellanies were again the issue, and the program director, furious at the announcing staff's disregard of my earlier memo, had issued one of his own, limiting the amount of rock during miscellanies to ten per cent. I also found, during my first days back, that the management would not be unhappy if I fired Charles for "irresponsibility," a term many on the staff considered a euphemism for insubordination.

For a few days I listened to Pitts during his announcing shifts, trying to decide whether to fire him or simply have another in our endless series of heart-to-heart discussions. Before I could make a decision, however, the program director heard a piece of rock music on a miscellany broadcast prior to the news. Enraged over what appeared to be (and admittedly was) a deliberate flaunting of his recent directive, he stormed into the control room and fired Pitts on the spot.

Pitts, along with those who felt he had been unjustly dismissed, called a staff meeting, at which he and a number of others expressed the opinion that he had been fired because of fears that the station and its programing were "being taken over by homosexuals"; that they were threatened not only by Charles' programs but also by his recruiting and training of other active gays for volunteer work around the station, and that the "rock miscellany" issue was merely a shield. But the decision to dismiss him was irreversible.

In another admirable display of persistence and determination—which he attributes to the realization that if it couldn't happen at BAI it positively wasn't going to happen anywhere else—Pitts again hung around the station as a volunteer. Several months later, under a new program director, I again rehired Charles, believing that his contributions to the station outweighed—barely—his apparent compulsion to destroy it, along with himself.

Two or so years of relatively good behavior later, Charles became the host and producer of *Out of the Slough*, a two-hour, Saturday-afternoon, free-form program aimed at New York's gay community. Attempting, in his on-air discussions of sado-masochism and pederasty, to push the barriers back beyond even legal and/or useful limits, Charles alienated as many listeners as he attracted. But those he did attract became his stanch supporters. At the end of June 1972, when *The Outside* died of old age after nearly seven years, Pitts, backed by his militant supporters, laid claim to the Saturday-midnight time slot. It was turned over to him on a three-month "trial" basis.

At the end of those three months Charles was removed from that slot. Management stated that the program dwelt on too narrow a range of subject matter for that particular time. His removal sparked a minor controversy among the gay community, and a major one within the station. New York's liberal weekly, *The Village Voice*, did a piece about it, followed, in *Voice* fashion, by several weeks of debate in the "Letters" section.

Meanwhile Pitts, in his fashion, took his case to the air, along the way publicly picking a number of musty Pacifica skeletons from the closet and holding them up for inspection by the entire listener-sponsored world. No one on staff who had had anything to do with the decision was spared Pitts's considerable personal wrath. Though I did not hear the program, accounts of it indicate that it was consistent with his past Pacifica history of excessive, occasionally destructive behavior. Still, management attempted to negotiate with him, even offering the same time slot on another night. But Pitts stood firm, citing the fact that Saturday night is the loneliest night of the week for the lonely homosexual—somehow more so in Pitts's peculiar perception than for the lonely heterosexual.

Finally, after months of routinely fruitless debate among the staff, and bowing at least partially to militant outside pressures (as has been the case too frequently in the recent history of WBAI), *Out of the Slough* was returned to the Saturday-midnight time slot.*

It is sometimes difficult to determine where political struggle ends and therapeutic need begins at WBAI. Still, the barriers do not come tumbling down even at WBAI without the exertion of a good deal of force. Charles Pitts's history at the station is only

---

* But, alas, not for long. During May 1973, as this manuscript was being laid to rest, Charles Pitts was fired by WBAI's current manager, Jerry Coffin. Although an explanation of the reasons for his most recent dismissal would require an additional chapter, albeit a dismally boring one, suffice it to say that the content of his program was *not* the issue. As of this moment (November 1973), Charles has not reappeared at the station. But tune in again at this same time next week. . . .

slightly more excessive than that of others, some of whom hold positions of power and responsibility within the institution today. The fact that WBAI has survived for over twelve years as an alternative form of communications, while most others have either ceased to be, or ceased to be an alternative, can at the very least be attributed to its rigid flexibility. The transitions have not always been smooth, but the station has somehow managed to keep contained, if at times just barely, those self-destructive seeds.

*The Pacifica Coat of Arms*

Logo...

# 9 *"We Are the Streets":*
## *All the News*
## *That's Fit to Be Tried*

At approximately seven-forty-five P.M. on Thursday, September 16, 1971, the phone rang in WBAI's master control room. The WBAI switchboard closes at six-thirty and all incoming calls are automatically transferred to "master." The announcer on duty answers the phones. It is a duty accepted and carried out with a sense of extreme indignation. As often as not, the phones go unanswered after six-thirty, though the fate of the universe might hang in the balance. And should the phone, by some monumental act of God, be answered, the party at the other end will be made to feel as though he or she has intruded upon the most intimate act of consenting adults. Which might actually be the case.

Neal Conan, acting news director that week while Paul Fischer was upstate covering the Attica Prison story, had finished the early-evening news broadcast about ten minutes earlier. The news that night had run more than twice its scheduled length, as had been the case for nearly a week. At WBAI, unlike other electronic newsgathering operations, content determines length. Now Conan was in the tiny news office stripping the wires, combing through newspapers and magazines, and updating copy for the late news and war summary.

Charles Pitts, the announcer on duty, answered the phone, probably expecting an agitated subscriber complaining for the sixth straight month about not having received the program guide.

The voice on the other end identified itself as a Weatherperson. "There is a communiqué from the Weather Underground for WBAI in a phone booth on the corner of ⸺ and ⸺," said the voice. "Send someone over to pick it up right away." Click.

Pitts passed the message to Conan, who has probably seen one too many B movies about newspaper city rooms, and without a moment's hesitation grabbed his trench coat and hat and dashed out the door. If he'd had even a moment's prior notice, he might have turned to the assembled few and announced, "This is the scoop we've been waiting for, boys!"

Less than ten minutes later Conan returned, out of breath, carrying a two-page typewritten message. The letterhead was a red lightning bolt cut diagonally by a black arrow, drawn in crayon. Those around the station who had seen previous Weather Underground communiqués said it looked authentic. (At which point I had a fleeting vision of a future singing Weathergram.)

This communiqué, like most of those that came before it, announced an imminent action. And also, like the earlier ones, it was skillfully designed to manipulate the media into disseminating its message.

It was dated September 16, 1971, and headed "Weather Underground Communiqué No. 11":

By now everyone is aware that the Monday slaughter did not have to take place. If this were a civilized society, the men in power would not need to kill those who demand their freedom and to be treated with the respect due to every human being.

Attica Prison is a place where 85% of those held there are Black and Puerto Rican. All the guards and administrators of the prison are white. This is not an oversight by some dumb bureaucrat. This is how a society run by white racists maintains its control.

Everyone knows about high bails, the box, beatings by white racist guards carrying "nigger sticks." Everyone saw Governor Rockefeller, Commissioner Oswald and the rest of those racists lie, and then attempt to justify their lies as to the alleged "killings" of the hostages.

It is not a question of being ignorant of the facts. In their manifesto the prisoners said: "The entire incident that has erupted here at Attica is a result of the unmitigated oppression wrought by the racist administration network of this prison. We are men, we are not beasts. And we do not intend to be beated [sic] or driven as such. What has happened here is but the sound before the fury of those who are oppressed."

Either you are a racist and support the murder and torture of Blacks and Puerto Ricans by mad-dog whites, or you commit yourself to doing everything and anything necessary to support the fight being waged by Black and Puerto Rican people in this country for their survival.

Mass murder is not unusual in this country: it is the foundation of Amerikan imperialism. In our lifetime we have seen four black girls killed by a bomb explosion in a church in Birmingham, Alabama. We have seen black students gunned down at Orangeburg, South Carolina and Jackson, Mississippi. We have seen Watts, Newark and Detroit. Amerika has murdered Malcolm X, Martin Luther King, Fred Hampton, and two weeks ago, the authorities in California assassinated George L. Jackson. We have seen white students shot and killed at Berkeley and Kent State [text missing, about one sentence] genocidal war will be more devastating than that waged by the Nazis—poisoning a people and a land for many generations to come.

Children grow up in this country knowing that Lt. William Calley can be convicted of the murder of 22 unarmed Vietnamese civilians and be congratulated for it by a President more interested in his re-election than the lives of any human beings on Earth. The main question white people have to face today is not the state of the economy (for many, the question of selling their second car), but whether they are going to continue to allow genocidal murder, in their name, of oppressed people in this country and around the world.

Tonight we attacked the head offices of the New York State Department of Corrections at Albany. Tomorrow thousands of people will demonstrate in New York and around the country against this racist slaughter. We must continue to make the Rockefellers, Oswalds, Reagans and Nixons pay for their crimes. We only wish we could do more to show the courageous prisoners at Attica,

*San Quentin and the other 20th century slave ships that they are not alone in their fight for the right to love.*

Conan raced into master control, interrupted the program on the air, recounted the phone call, and read the communiqué. Then he went back to the news office, and he and Margot Adler (the only two newspeople around the station at the time) tried to reach sources in Albany to confirm the announced action.

Meanwhile telephones began to ring all over the station. Other news-gathering organizations were calling—TV networks, the wire services, local radio and TV stations. Despite their experienced news staffs, millions of dollars in resources and equipment and "reliable sources," they had been scooped by that boil on the FM band, WBAI.

At 8:01 P.M. a bomb exploded near the office of State Corrections Commissioner Russell Oswald. There were no injuries.

By that time Paul Fischer and Bruce Soloway, who had been covering the Attica story for the past week, were on the scene. They phoned the station and did a live, firsthand broadcast. At about that time most other correspondents were, most likely, impatiently awaiting an official press release from the Albany Police Department.

And so it had been all week. Those who were genuinely interested in knowing what was happening inside and outside the prison knew where to tune. Not to CBS, NBC, or ABC—with their extraordinary resources and skill, but to WBAI, with a news staff of six, average age about twenty-one, little if any professional training, and a budget for the operation of the entire News Department far less than Walter Cronkite's annual salary.

The previous Saturday, when the situation at Attica seemed to be growing more tense, Paul Fischer had hopped a plane headed for Albany. Before he left New York he phoned ahead to reserve a car from Avis, the folks who try harder, in the name of Pacifica Foundation. The reservation was accepted and confirmed with synthetic graciousness—which abruptly turned to genuine alarm the moment the Avis representatives set eyes on WBAI's news director, his shoulder-length hair pulled back in a pony tail, his person packaged in jeans and a workshirt. This may have prompted them to do a quick telephone credit check on the station, which in turn seemed to convince them that they had never made such a long-distance commitment.

Fischer phoned the station, pissed to a falsetto, and urged Ed Goodman, WBAI's station manager, to use the influence and prestige of his good offices. Which, as usual, got us nowhere.

By then it was ten P.M. Fischer had been at the Albany Airport for more than two hours, trying to get the folks who try harder to try *harder*. And so, at a few minutes after ten, Neal Conan went on the air with an update of the Attica story, casually mentioning at the end that he had hoped to have a live feed from Fischer at the prison, but he'd been unable to get there because Avis refused to release the reserved car. As a public service Conan offered the telephone number of Avis's Albany offices.

Half an hour later Fischer was on his way in a shiny new Dodge Swinger. The telephone lines at Avis's Albany offices remained tied up for two hours, at which time they phoned WBAI and begged that we call off the dogs. It is a modest example of the too infrequently used power of the electronic media.

---

*International Hotel*
*6 Donegall Square South*
*Belfast, 3T15JR.*
*Northern Ireland Apr. 16, 1972*

TO: *Steve Milhous Post*
*c/o WBAI Radio*
*359 E. 62nd Street*
*New York, New York 10021*
*U.S.A.*

TOP SECRET SENSITIVE:
EYES ONLY FOR THE PRESS

*Dear Steve Post,*

*I know I promised to send a scenic postcard but I decided against it for 2 reasons:*

a) *there are no scenic postcards (except for the ones which seem to have had the words Palm Beach crossed out in the bottom corner);*

b) it's too easy for the Special Branch to read a postcard; this way, they have to take the trouble to steam open the letter and reseal it—that being such a complicated task, that they may not bother.

And now the headlines:

FISCHER AND CONAN ARRIVE IN BELFAST, STEP OUTSIDE THEIR HOTEL, AND A BOMB GOES OFF. . . .

FISCHER AND CONAN GO OUT TO BUY A NEWSPAPER AND THE EUROPA HOTEL BLOWS UP. . . .

FISCHER AND CONAN TAKE A WALK DOWN DONEGALL STREET AND A BOMB GOES OFF AT IVAN AGNEW'S MERCEDES-BENZ AUTO SHOW-ROOM. . . .

FISCHER AND CONAN, FEELING A BIT LIKE ALBATROSSES, REPAIR TO THE NEAREST BAR AND GET "BOMBED" THEMSELVES. . . .

FISCHER AND CONAN DECIDE TO TAKE A WALK DOWN FALLS ROAD. UNFORTUNATELY THEY DECIDE TO DO SO AS BRITISH RED BERET PARA-TROOPERS HAVE DECIDED TO SET UP A PATROL THERE. . . . AS A RESULT, FISCHER AND CONAN ARE SEARCHED AT GUNPOINT. . . .

FISCHER TAKES A SATURDAY-NIGHT WALK AND ENDS UP BEING MA-ROONED BEHIND A BILLBOARD ADVERTISEMENT FOR THE SALVATION ARMY, AS CHILDREN AND BRITISH TROOPS AMUSE THEMSELVES BY TOSS-ING STONES AND FIRING RUBBER BULLETS AT EACH OTHER (APPAR-ENTLY, THE LOCAL VERSION OF JOHNNY-ON-THE-PONY PLAYED WITH ARMORED CARS AND RIFLES). . . .

FISCHER GOES BACK TO HOTEL AND TRIES TO GO TO SLEEP BY COUNTING SHEEP; INSTEAD, ENDS UP DROPPING OFF TO SLEEP BY COUNTING RIFLE SHOTS AROUND THE CITY. . . .

FISCHER CONSOLES HIMSELF WITH THE THOUGHT THAT NO MATTER WHAT, IT'S STILL MORE CIVILIZED THAN A BAI STAFF MEETING (AT LEAST AROUND HERE, YOU KNOW WHO IS SHOOTING AT WHOM). . . .

And now, details on these and other stories:

Fischer and Conan (Conan, by the way, has since departed for Londonderry) were walking down Castle Street leading to Falls Road when they happened to pass a parking lot. It was your average run-of-the-mill parking lot, but it also contained (unbeknownst to us) your average run-of-the-mill contingent of British Army troopers complete with combat uniforms, truncheons, M-16 rifles, and an 8 × 10 color glossy picture of the Queen. As we passed, we were "invited" inside the parking lot. Being the innocent chap that I am, I assumed that we would be offered a cup of English Breakfast Tea, but instead we were invited to be the recipients of an up-against-the-wall spread-eagle search of our personages, at gunpoint. As the Queen herself would say: We were not amused. Many strange and wondrous thoughts pass thru your mind at such moments, like:

1. I wonder if that M-16 is loaded?
2. How does "God Save the Queen" go?
3. Hey soldier, wanna buy some French postcards?
4. The American ambassador shall hear of this!
5. If I sneeze, will they think I've insulted them in Gaelic?
6. Hey you guys, we helped you win the last war, didn't we?
7. I voted for Twiggy in the Top Model contest.

Anyway, such as it was, we lived thru this ordeal at the hands of Her Majesty's Own (her majesty's own what, I don't know, but if she owns them, somebody ought to tell her what they've been up to lately). . . .

There isn't much to do on a Saturday night in Belfast—most of the moviehouses are closed due to bomb damage, and the only one that was open was playing Chitty Chitty Bang Bang, which was a pretty good description of what was going on outside the theater, as opposed to inside. Anyway, I decided to go for a walk, my rationale being that I'm neutral in all this—which, in practical terms, means that I have an equal chance of being shot by the Catholics, the Protestants, the Royal Ulster Constabulary, the British Army, the IRA, the provisional IRA, the special branch, District Attorney Hogan, the CIA, the KGB, and other sources. Now under these circumstances, I could be shot and killed by all of them (one of the unique features of a documentary) or, if they all missed, there was a good possibility that they would shoot each other. As such, my chances for surviving the walk were 50-50, and around here, that's better than average. As I walked down the Falls Road, I noticed a group of young children diverting traffic. I thought to myself: Aren't these youngsters wonderful. Directing traffic, because the traffic signal is broken. As I rounded the bend, it came to my attention that there were 5 British armored cars on the street, that soldiers had taken up positions behind them, and their rifles and guns were aimed in my general direction, and the children in front of me were bombarding the troops with rocks and bottles. As my life flashed before me (I thought I would at least be spared the agony of this year's marathon) the troops fired a rubber bullet, which fell five feet short of my right buttock. Not having a white flag with me tonight, I took refuge behind a billboard advertising Ex-Lax, which under these conditions, was just about the last thing I needed. . . . Anyway, having a wonderful time, Wish You Were Here (instead of me).

FISCHER

---

In the course of gathering news, WBAI has, occasionally, made news. This is not an entirely surprising occurrence, since, in order to maintain an effective, meaningful news-gathering organization with our tiny budget and journalistically inexperienced staff, we have had to rely on sources either shunned, or considered of second-fry value to the major news-gathering organizations. It would be folly for WBAI to attempt to compete with CBS or *The New York Times*, and so it is all the more extraordinary that the station, with its meager resources, has captured a major share of the New York City FM audience (according to the ratings) for both its early-evening newscast and the war summary, a WBAI institution since 1967. More surprising, the WBAI news department, under the direction of Paul Fischer, has gained the grudging respect, admiration, and ear of every serious journalist in the New York area, as well as a number of the industry's most coveted awards (including the 1972 Overseas Press Club Award, for Paul Fischer and Neal Conan's three-hour documentary on Northern Ireland, entitled "A Month of Bloody Sundays"). Fischer, typical of the "Pacifica Orphan" syndrome discussed earlier, very nearly literally stumbled into WBAI as a volunteer in 1966, at the pliable if not tender age of eighteen. After falling in and rapidly out of a number of areas of programing (including two months of doing *The Outside* during my absence, which he considers the most horrifying of all his Pacifica adventures), he found his niche in the news department.

The day-to-day operation which produces WBAI's six-thirty P.M. news remains, after more than eight years, somewhat of a mystery to me, and, I suspect, to most of the staff. When I first arrived at WBAI, our daily fifteen-minute newscast, though much lauded by the audience, was little more than a rewrite job of the major wire-service stories, unloading, in WBAI fashion, such phrases as "Communist terrorists" and substituting in their place "forces of the National Liberation Front." WBAI began questioning, long before the established news vehicles did, the practice of printing government press releases verbatim and without query as to their accuracy. (For example, Vietnam casualty figures: apparently it wasn't until some researcher with an adding machine and an almanac deduced that, according to United States Government sources, we had already killed the entire population of North Vietnam several hundred times over that the press began qualifying the government reports by stating "enemy casualties were *listed* at . . .")

In the late 1960s, it was recognized that information from Communist countries which excluded American correspondents was becoming increasingly crucial to a complete understanding of world affairs. Fischer's predecessor, Paul Schaffer, believed that WBAI could greatly expand its coverage of those areas with the addition of Agence France Presse, the French news service, which maintained correspondents in Hanoi and Peking, among other places not covered by AP and UPI. But, typically, there was simply no money available for it. So Schaffer offered to pay a portion of the expense from his own meager salary; at the same time he went on the air at the end of each day's newscast to ask those listeners who appreciated the expanded coverage to send in special contributions earmarked for the payment of AFP. Before long—though not before the agency had sent several notices threatening to terminate its service due to nonpayment of bills—the additional contributions were paying for AFP. Later WBAI added Reuters, the British news wire, and began subscribing to hundreds of journals, obscure and well known, politically left and right, from the far corners of the world. And finally, again in WBAI fashion, we installed a special "news hotline" telephone, the number of which is given out daily over the air for listeners to call in "possible news items." Though more often than not these tips turn out to be useless, occasionally they have uncovered stories not found elsewhere. Today, one can only assume, Fischer maintains and cultivates a rather well-informed, unconnected unofficial network of "anonymous sources." It is the maintenance of the anonymity of these sources which has led WBAI to being, as well as gathering, the news.

Some, in fact, maintain that making the news is WBAI's role. This idea was best articulated by Paul Gorman and Nanette Rainone in a piece entitled " 'Phonecasts' and the Right of Public Assembly," which was printed in WBAI's *Folio* in May 1972:

> . . . *it must be said that BAI is not just a news organization, reporting other people's activities. We are, and have been for some time, a new form of neighborhood—not just an institution supported by the community, but the community itself—the people in microcosm. By combining the economic model of listener-support with the electronic mode of the "Phonecast," our community has created for itself an entirely new forum for public dialogue, an electronic Town Hall permanently available to the people for their own purposes. So when women are discussing women's issues on the air, we are not reporting an event, we are an event—another meeting of the community through a "Phonecast." When teachers, parents, and students are discussing the school strike, when members of the Forest Hills community discuss living with low-income housing; when war resisters discuss war resistance; when junkies discuss junk; when homosexuals discuss homosexuality; and when prisoners and jailors discuss jail—we are not just a news organization; there is no media-tion, no out-takes. We are a publicly-financed public instrument for public discourse. We are, in effect, the streets. . . .*

Over the years, WBAI's audience has come to expect this kind of coverage from WBAI—coverage that goes beyond a mere statement of the "facts." When there is a significant event occurring in the New York area—or elsewhere, for that matter—it has become routine for WBAI to pre-empt its regular programing and cover, in as much detail as possible, that event. For example, WBAI was doing live, complete coverage of the major Vietnam-war demonstrations, both anti and pro, while major broadcast news operations, including "public" television, were still devoting three minutes on their nightly newscasts to a "balanced" representation of those activities. Many, in fact, credit WBAI's initiative as the prime force in the networks' live coverage of these annual events in the late 1960s.

(During one such event, which took place on a Saturday afternoon in April, a WBAI listener phoned to tell us that she had called WNET, New York City's public TV outlet, to ask why they were not doing extended coverage of the event. "Our news department is closed on Saturday," was the matter-of-fact reply, she said.)

It is such coverage that led to "The Great WBAI Tape Controversy" of 1971–1972, a case which, according to WBAI's attorneys, raised (though never solved) "fundamental questions of the media's right to report and the people's right to speak and the people's right to hear and know."

It began in October 1970, during the insurrection at the Men's House of Detention (the Tombs) in New York City. Tombs inmates had taken a group of guards as hostages, in an attempt to force the Department of Corrections to respond to the hideous conditions at the prison, where a good number of the inmates, mostly poor blacks and Puerto Ricans were (and are) serving time without benefit of trial, often with no bail, or bail set so high it is far beyond their means. All this is the result of an ancient, inept, inefficient, and often corrupt judicial system.

Typically, WBAI had been devoting a good deal of coverage to these conditions prior to the insurrection, and as a result was able to make contacts within the prison during the rebellion. And so, between October 2 and October 5, WBAI broadcast nearly thirty-five hours of material relating to the rebellion, including many hours of live, on-the-air phone calls with both the prisoners and their hostages.

The "Tombs Riot," as the established press dubbed it, came to an end with Mayor John V. Lindsay's promise of amnesty for the participants, as well as a promise of redress of the prisoners' major grievances. None of the hostages had been physically harmed and, in fact, many spoke later of the humane treatment afforded them by the prisoners, as well as their basic agreement with the prisoners on the need for reform.

But before the dust could settle in their cells, New York's Dis-

Paul Fischer delivers news
of the day's disasters in
the old 39th Street studio,
circa 1968.
(PHOTO: STEVE POST)

The Pacifica "team" (Dale Minor, Peter Zanger, Steve Post) covering
the anti-war moratorium held in Central Park, New York, April 1968.
(PHOTO: PAUL BUSBY)

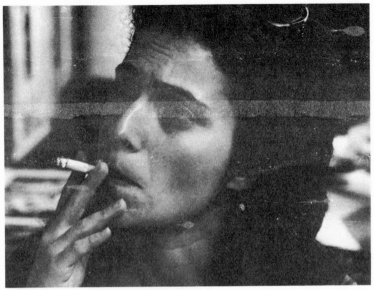

*Above:* Nanette Rainone, WBAI's first feminist program director. She originated on-the-air "consciousness-raising" and co-produced, with the author, *The Sex Programme. Below:* Ed Goodman, WBAI's former station manager. (Ed is now under heavy sedation.)
(PHOTOS: © JIM DEMETROPOULOS, 1974)

trict Attorney Frank Hogan, an institution in the city (the same D.A. who nearly prosecuted Lenny Bruce to death), announced that he would press charges against a number of the prisoners involved in the Tombs rebellion, despite the Mayor's promise of no reprisals.

"It was against this background," wrote WBAI's attorneys, Rhonda Copelon Schoenbrod, Jeremiah Gutman, and Peter Weiss, in the May 1972 *Folio*,

*that Detective McCarthy, of District Attorney Hogan's office, rang the bell at WBAI's offices on January 28, 1971, and served upon Ed Goodman, in his capacity as the station's general manager, a subpoena, duces tecum, which is an order, drafted by a lawyer— often a prosecutor—requiring a person to appear at a certain time and place, under penalty (sub poena) of law, and to bring with him or her (duces tecum) certain documents or other materials. Here Ed Goodman was to appear in the criminal court a full two days later, supposedly at a trial which, in fact, was far from even the picking of a jury, and bring with him: "all tapes, recordings, program logs, and material broadcast with relation to the riot at the Manhattan House of Detention for Men (Tombs) during the period between October 2, 1970, and October 5, 1970."*

It was not the first subpoena to be served on a news-gathering organization, nor indeed the first to be served upon WBAI, but it certainly was the broadest—calling for the better part of three days' worth of broadcasts and related material. And there was no attempt to conceal the use to which the D.A. planned to put the tapes: they were to be used as evidence—more than likely to identify the suspects through voice prints—against the "Tombs 7." *WBAI had been ordered to become an investigative arm of the District Attorney's office.*

Initially, WBAI asked for a two-day extension of the deadline in order to decide whether or not to comply. But the request was denied. WBAI's attorneys described what happened then:

*So, on the 31st (of January), WBAI and lawyers from the Center for Constitutional Rights met with John Fine, the Assistant D.A. responsible for this subpoena, at which time he graciously took the* position that all he wanted was material arguably relevant to the events in question, which was actually broadcast from the time of these events through the Spring of 1971.

At this meeting, it was decided to turn over to the D.A. all program logs (brief written descriptions of aired programs) relating to programs dealing with prison conditions over a period covering six months, on the theory that such logs, being required by law to be kept and made available to one government agency, the F.C.C., could not reasonably be withheld from another.

There was one major difference between this and other cases in which reporters' and news-gathering organizations' materials had been subpoenaed: most of the tapes ordered to be turned over to the D.A. had already been broadcast, thereby, in the opinion of the prosecutor, giving them the equivalent status of already published material. Still, in the opinion of WBAI and our attorneys, it was "privileged" material; after all, the prisoners and hostages who had spoken over WBAI's airwaves during the days of October 1970 had done so anonymously. Had it been made clear to them that their words and voices would be turned over to the D.A.'s office as evidence in their own prosecution, they more than likely would not have consented to the broadcasts.

In addition, article 79-h of the Civil Rights Law of New York, the state statute under which WBAI claimed immunity, makes no distinction between broadcast and nonbroadcast materials.

WBAI's attorneys described what happened in court on Monday, February 28, 1972:

. . . a motion was made on WBAI's behalf to quash the subpoena on several grounds, including violation of the First (Freedom of Speech) and Fourth (Search and Seizure) and Fourteenth (Due Process) Amendment and Article 79-h of the Civil Rights Law of New York which, on its face, exempts all newspeople, including broadcasters, from subpoenas. At the conclusion of the argument, the judge announced that he had no problem whatsoever with this subpoena and that if people don't want their statements used against them, they shouldn't make them, saying, "If anybody wants

to keep news to himself, he should keep his mouth shut, or his
typewriter don't work [sic]."

The station having refused to turn over the materials in question, the D.A. obtained an order to show cause why WBAI and
Ed Goodman should not be held in contempt. WBAI's attorneys
asked for a stay of the contempt proceedings, pending appeal of
the subpoena in the Appellate Division. On March 2, Justice
Gerald Culkin denied the motion.

WBAI was ordered to return to court the next day to comply with
the subpoena or be held in contempt. Prior to that appearance, an
application for an emergency stay of any contempt proceedings
was made to and denied by a single Justice of the Appellate Division, but set down for consideration by a panel of five judges of that
court the following week.

At the Friday hearing in the Supreme Court, Justice Culkin in-
voked a power reserved by law for those rare occasions of imminent
and actual disruption of judicial proceedings, to hold WBAI and
Ed Goodman summarily in contempt. No hearing was had, no witnesses were permitted to testify, not even the minimal showing of
need for the materials subpoened was required from the D.A. and
Ed Goodman was denied the most fundamental right to address
the court. Rather, the station was held in contempt and fined $250
and Ed Goodman sentenced to 30 days in the civil jail. Motions
to stay execution of the sentence or release him on bail pending
appeal were rejected. This last refusal appears to be unprecedented:
reporters and other newspeople have gone to jail in the past, but
not until their legal claims were finally determined and rejected on
appeal [italics mine].

WBAI, now perhaps for the first time realizing the extent of the
D.A.'s determination, and fearing the issuance of a search warrant
and confiscation of the tapes, took new precautions. What limited
taped records existed of the Tombs broadcasts were removed to an
undisclosed location, far from the vulnerable halls of the studio,
to insure that they would not mysteriously disappear. Further,
entrance to the WBAI building was severely limited. No one was

permitted in without some sort of identification, and all inside were instructed to refuse admittance to the police, unless they could show a proper search warrant.

On Friday, March 3, at about four P.M., Ed Goodman left his humble Park Avenue dwelling and was led off to the civil jail on Manhattan's West Thirty-seventh Street. He must have been seriously re-evaluating his decision to disown the family business and leave a potential life of leisure. Ed Goodman was in the clink.

WBAI, never one to learn from the past, interrupted its regular programing to provide full coverage of the issues, doing live, on-the-air interviews with members of the press and others (such as some of the prisoners and former prisoners who had been involved in the Tombs rebellion) who had an interest in the case.

Again, WBAI's attorneys described the next judicial procedures:

A second application, this time for a stay of the sentence, was made to the same Appellate Division Justice at an extraordinary Saturday morning session and was similarly denied, but set to be heard the following Tuesday.

With no relief in sight in the interim, Center [for Constitutional Rights] lawyers, joined and bolstered by the New York Civil Liberties Union, determined to take the matter to the federal court which has the authority to issue the "Freedom Writ" (writ of habeas corpus) to correct flagrant injustice. So, on the morning of Sunday, March 5, United States District Court Judge Marvin Frankel, having determined that substantial rights and issues were at stake, did, in an unusual session convened near his residence in the Larchmont Village Hall, sign a writ of habeas corpus, which ordered that Goodman be liberated forthwith. In so doing, Judge Frankel noted:

If I were to be too unguarded on this Sunday morning, I might take judicial notice that he (Goodman) is manager of a radio station which is not committed always to orthodoxy, and that may have a bearing in this case, too, but I think at most a peripheral one.

The tide had turned.

On the following day, Mr. Hogan—now represented by another

District Attorney, filed an appeal from Judge Frankel's order to the Federal Court of Appeals for the Second Circuit, seeking to have it determined that the federal courts had no business redressing such injustice, no matter how grave, done by the state courts, even though a footnote in his appeal brief indicated that he was no longer interested in Goodman remaining incarcerated while his case proceeded through the state courts.

On Thursday, March 9, in one of the more bizarre incidents in recent judicial history, three judges of the Federal Court of Appeals were hearing arguments on the D.A.'s appeal from Judge Frankel's order at the same time that five judges of the New York Appellate Division were debating Goodman's applications for a stay or bail. By late afternoon, three of the latter had declared the issue academic, in view of Judge Frankel's intervening order, while a minority of two held that, since substantial questions had been raised by the motion to quash the subpoena, the stay should be granted.

The Court of Appeals subsequently reversed Judge Frankel and ordered the writ of habeas corpus vacated, not on the broad jurisdictional ground urged by the D.A., but on the narrowest ground, that Judge Frankel acted too precipitously, saying in effect that for Ed Goodman to be incarcerated from Sunday to Tuesday, the time scheduled for the hearing before the Appellate Division, would not have been such a big deal as to require federal intervention. But, the Second Circuit stayed their order to permit a new application to the Appellate Division which immediately granted the stay, pending Goodman's and WBAI's appeal on the merits.

Meanwhile, back at the station, while all these complex legal maneuverings were going on, the staff was running around a bit like the proverbial chicken with its head cut off. Our head was, after all, incarcerated—ironically, for exposing to the public a bit of the truth about the very judicial system he was now a victim of. While we may have seen it as ironic, Mr. Hogan, I suspect, viewed it as a piece of poetic, as well as judicial, justice.

The staff gathered together in what I consider one of the more curious phenomena of Pacifica, and the one, perhaps more than any other, responsible for the survival of the institution: the instinct to band together in moments of external threat. Suddenly it no longer mattered who stood where on what, who was fucking whom, or any of the thousands of other issues that threaten daily to destroy the station from within. All were united, exploiters and exploited, revisionists and revolutionaries—those who the day, even the hour, before had threatened to spit up at the sight of their mortal enemy. The persecuted stood together proudly before the firing squad. Alas, it would not be long before things were back to normal.

At any rate, since WBAI's defense was being handled gratis by the Center for Constitutional Rights and the New York Civil Liberties Union, it became the staff's task to take care of all the necessary legal and other paperwork. The lawyers prepared an *amicus curiae* (friend of the court) brief, which within forty-eight hours had been signed by more than 575 journalists. The brief did not attempt to base the case on its merits, since there was considerable difference of opinion on the issues, primarily over the fact that much of the material in question had been previously aired. But there was almost unanimous outrage over Ed Goodman's precipitous jailing. Much to the surprise of everyone on staff, newspeople of widely divergent backgrounds and political philosophies signed their names to copies of the brief. A small sample of that list includes the following:

*Gabe Pressman, reporter, WNEW-TV News; Eleanor Fischer, reporter, Canadian Broadcasting Corporation; Peter Arnett, Special Correspondent, Associated Press; Jimmy Breslin, author, free-lance journalist; Earl Caldwell, National Correspondent, The New York Times; Roger Grimsby, Anchorman, Eyewitness News, WABC-TV; Peter Gwynne, Associate Editor, Newsweek; Pete Hamill, columnist, New York Post; A. M. Rosenthal, Managing Editor, The New York Times; Paul Cleveland, News Director, WOR-TV; Max Lerner, columnist, New York Post and Los Angeles Times Syndicate; Sandie North, writer, Ladies' Home Journal; David Sherman, Senior Editor, Life; Les Brown, TV-Radio Editor, Variety; Ward Chamberlain, Executive Vice-President, WNET; Barry Farber,*

talk-show host, WOR-AM; Joseph Morgenstern, journalist, News-
week; Jack Nessel, Managing Editor, New York Magazine; Martin
Nolan, Washington Bureau Chief, Boston Globe; Gloria Steinem,
Editor, Ms.

The dichotomy of view regarding the nature of already broadcast
material was summed up best by a *New York Times* editorial of
March 6, 1972, entitled "The Goodman Jailing." In part, it said:

No defensible judicial purpose was served by the refusal of the state
courts to fix bail for Edwin A. Goodman. . . . In this state, the
broad guarantees of press freedom contained in the First Amendment to the United States Constitution are supplemented by a statute explicitly prohibiting contempt judgment for "refusing or fail-
ing to disclose news or the source of any such news." Unlike many
approaches to press freedom, which draw an artificial line of distinction between print and electronic journalism, the New York
statute protects newscasters as well as newspaper reporters and
editors.
   Our belief that there never was any warrant for court denial of
bail or parole to Mr. Goodman is not meant to imply a belief that
the WBAI case is necessarily sound. The basic question of infringe-
ment on the freedom of the press seems less than totally clear-cut.
It is certainly debatable that a station, which offers its tapes for sale
to anyone who is willing to pay the fee, is on solid ground in refusing to sell them to the District Attorney or in defying a court
order to make them available. Also decidedly debatable is the
WBAI contention that comments in a broadcast are not matters of
public record but are "public only for the fleeting moment during
which they are actually broadcast."
   Yet the very fact that there are still not trustworthy judicial
guideposts by which to operate in this field would have made it
grossly unfair to withhold from Mr. Goodman the same access to
bail pending appeal that is the right of persons found guilty of
major crimes of violence or betrayal of trust.

In a letter dated March 6, 1972, and published on March 15,
Frank Millspaugh, former WBAI manager, took issue with the

*Times* editorial, and in so doing best summed up the station's position:

. . . *While you correctly assess the indefensibility of the court in jailing station manager Edwin A. Goodman . . . your editorial dwells too heavily on legal distinctions of dubious applicability.*

*At issue here is a question of major public policy: whether the police function of government may coerce the press into performing an auxiliary role. Essentially, the District Attorney is attempting to obtain journalistic materials for use in criminal prosecution. If he succeeds, every newsman becomes a potential witness for the prosecution. What could produce a more "chilling effect" upon the First Amendment?*

*State law prohibits the court from finding in contempt any member of the press who refuses to reveal news or news sources. This is not because our legislators just enjoy inconveniencing the police; it is because they recognize that any compromise of the freedom and integrity of the press must result in greater social harm than that compromise could prevent. Securing additional criminal convictions is not worth the sacrifice of a free and independent press.*

*Further, the present technology of voice-printing from tape recordings poses special problems for the electronic media. Surrendering a tape of a man's voice is the equivalent of surrendering his thumbprint or his signature.*

*In our time, the press is the eyes and the ears of the public. These public facilities have suffered impairment from both weakness within and injury from without. I will not attempt to diagnose the pathology of your editorial myopia in this issue; I will call upon you to consider whether the subordination of press rights to police functions is not an essential attribute of the police state.*

During Ed Goodman's incarceration it was reliably rumored that the District Attorney's office, cognizant of the validity of the "fishing expedition" aspect of the case for the defense, planned to issue a second, narrower subpoena, calling only for materials which had actually been broadcast, with the intention of obtaining only those tapes containing the voices of prisoners. It was also becoming clear to them that such intimidating tactics as Goodman's jailing were not only not going to cause WBAI to bow, but would add new resolution to WBAI's resistance.

Now this gave a new dimension to the predicament: because it had all happened so rapidly and unexpectedly, no one had bothered to listen to the tapes in question. We had taken all the tapes from the station's archives with labels indicating any connection to the Tombs rebellion, hastily thrown them into a sack, and carted them off. Now they were secretly returned to the station, where three or four staff people under the supervision of our attorneys spent a good part of the day locked in control rooms listening to the tapes. There was among them, curiously, not a single air check (tape made from live broadcast) containing the voice of a prisoner calling from within the Tombs.

In the end the case was dismissed—but not because there were no tapes of the prisoners' voices; the court ruled instead that, as WBAI's attorney had argued, the District Attorney's subpoena was so broad and so vague as to constitute a fishing expedition. There were no legal precedents set with the case, perhaps, but at least WBAI was not forced to become an investigative arm of the government.

---

## NOTES FROM THE AUTHOR'S DIARY

TUESDAY, MARCH 7, 1972
NEW YORK CITY

. . . So Fischer and I had to leave and lug the entire sack of tapes back to the city less than twelve hours later. So much for our pleasant four days in the country. (Of course we never do anything but talk about the station and bemoan our fate anyway.) It's all too fucking much. I don't know if I'd have taken the job if they told me it included typing legal briefs, schlepping petitions around, sneaking past New York Times security guards at four in the morning, and smuggling tapes in and out of the city.

Nanette was really pissed. We missed her by about two minutes. If we hadn't gotten stopped by that schmuck cop upstate we would have made it on time. Of course, that didn't make her any more sympathetic. Thank God for Fischer's press card—that asshole might have shot us on the spot! Fischer looks like a fucking madman with his hair like that. If that cop had searched the car and found that mailbag full of tapes, he might have gotten suspicious. He probably would have arrested us for illegal possession of a U.S. Government mailbag. . . .

Anway, we made it to court just in time. Of course, nothing happened—all they did was send it to another fucking court. So Goodman will have another couple of days of martyrdom at least. . . .

FRIDAY, MARCH 10, 1972
NEW YORK CITY

They finally sprang Goodman yesterday. Practically the whole staff was there by the jailhouse steps trying to get their pictures taken for the tube with their returning hero. They acted like Perón was returning from exile. Naturally we got there just as he was getting into a cab.

Of course now Goodman's got the staff in his hip pocket. What a bunch of assholes! Break your ass to keep the place alive and they dump shit all over you. But spend two days in a plush civil jail defending "the cause" and suddenly you're the leader of "the people," head of the provisional revolutionary government.

Anyway, Goodman's really into it. At least he's playing along. He said it was a "radicalizing" experience. He'll probably wind up publishing his prison diary, all three days' worth.

Suppose I'm more than a little jealous. Martyrdom seems most attractive right now. . . .

The Bergdorf Man

* Studded cuffs    $3000.
* PLATINUM BARS    $8500.
* GIANT MONOPOLY SET   $2000.
* Bail    $25000.

GO TO JAIL

LEVENWORTH / PARK PLACE

# 10  Son of Playlist: The Decline and Fall of Commercial Free-Form Radio

A playlist is a list of records or record cuts to be aired in any given hour of radio broadcasting. It also gives the order in which they are to be played and indicates what commercials or public-service announcements (PSAs) are to be broadcast at what times.

The playlist is produced by the program director, who must keep in mind the demands of the advertising and continuity departments. Together they see to it that you, the radio listener, are not exposed to any undue creativity over the airwaves. It's nice to know you're being looked after.

Playlists also insure that all the major record companies will get their share of air play (i.e., "plugs") for their latest releases. ("Biggest thing since the Beatles," was the typical hype of greasified promotion men during the late 1960s. It hardly mattered whether the record was a Jackie Gleason–style instrumental or a new group manufactured in the bowels of the record company.) Playlists also guarantee that you won't be offended by *anything* heard over their air—including the news. Most radio stations subscribe to either "rip-and-read" news wire services, or to the similarly prepared, prepackaged audio news service. Thus the news becomes another Saran-wrapped, standardized commodity, pretenderized and easily digestible. Just like the product that's probably sponsoring it. If you should happen to be offended by something you've heard on the news—offended enough to complain to the station— you are likely to get an apology and the phone number of the news

service. The station's hands are clean. They haven't offended anyone—they never would—on pain of losing an advertising dollar.

But let's go back a minute to those playlists. As a listener, I had
always thought that the guy who was talking to me on the air was
also picking out the records. Not that it was something I spent a
lot of time thinking about—I simply took it for granted. It didn't
seem as if it should be too difficult a job for one person to handle.
In fact, I worked on the air at WBAI for almost two years before
I even learned what a playlist was.

It happened one day about six years ago while I was visiting the
studios of WNEW-AM, a "middle-of-the-road," financially successful station in New York City. (WNEW-AM is the station responsible for making *Martin Block's Make-Believe Ballroom* a house-
wife's habit in New York during the fifties. Many women probably attained an active and satisfying sexual fantasy life during
those years listening to Martin's voice while accomplishing their
dreary daily chores. They never suspected for a moment that Martin was a fat, bald little man who couldn't have sold a used car on
TV.)

I talked there to Ted Brown, then one of WNEW's highly paid
disk jockeys, during a five-minute news break. He complained about
the shit he got from the programing people every time they caught
him slipping in a record of his own choosing. I was really amazed.
Here was a man working on the radio three hours a day, five days
a week, making almost a hundred thousand dollars a year, and he
wasn't even given the responsibility of choosing his own records.

Just what is it that this man does for a living that couldn't be
done just as well, and probably more cheaply, by a computer? I
pondered. The answer, as I see it, is nothing. But, so far, union
contracts forbid the use of machines as air personalities. Who says
labor unions have lost their humanity?

And so, as I discovered, playlists have nearly always been standard operating procedure at almost all radio stations. As an
excuse for their objectionably bland programing, their playlists,
and their rip-and-read newscasts, station managements will often
cite the Federal Communications Commission's references to "the
licensee's responsibility" to remain aware at all times of the content

of the station's programing. Unfortunately, it seems that in general, radio-station managers are unable to differentiate between *having knowledge of* and *exercising control over* program content. Management has always chosen to interpret FCC rules in the narrowest possible context, thereby insuring the continued harmlessness of your radio.

However, a few years back, as a concession to the "now generation," some radio stations began to abandon playlists and gave at least musical control to the person on the air. Their aim was to produce a younger, more spontaneous and human sound. Anyway, that was the image they sought. For the most part it amounted to fitting Murray the K with a shag hairpiece, bell-bottoms, and an updated vocabulary of hip expressions.

But there were some serious attempts, too. It all started in the early 1960s on WBAI, with Bob Fass's *Radio Unnameable*. But of course nobody paid any attention to it at the time because WBAI is licensed as a "noncommercial, educational station." And if it doesn't sell soap . . .

By the time WBAI began to show up in the FM ratings, *Radio Unnameable* had been on the air for a number of years, and had been joined by a couple of other similar programs on WBAI. Some of our early listeners had gone off to college, joined the staffs of their college stations, and in a few cases managed to change the entire format of their stations to the Fass-like free-form.

The most successful of these and the one I believe credited with originating the term "free-form," was WFMU-FM, the radio station of obscure Upsala College, in East Orange, New Jersey. (WFMU, like a few other experimental college radio stations, eventually went back to playlists and rip-and-read news when one day someone from the administration happened, accidentally, to turn on the station. Most of the staff walked out rather than return to the old format.)

When the Big Boys in broadcasting began to see that WBAI in New York and WFMU in New Jersey (along with a few others around the country) were getting respectable shares of the FM rating pie, they suddenly developed a deep sense of commitment to "the youth." They renamed it the "youth market."

The people at WNEW-FM, Metromedia's FM outlet in New York, called their format "The New Groove." They rescued both Scott Muni (a relic of screaming 1950s AM radio) from Jingle Village (a retirement community for the eternally raspy), and Roscoe (in real life Bill Mercer) from the shambles of the WOR-FM free-form experimental format, leaving Murray the K unemployed—with his new hairpiece, bell-bottoms, and vocabulary. To round out their new youth sound, WNEW hired prepubescent-voiced Jonathan Schwartz from Boston, and kept Allison Steele, "the night bird" who had been with the station through its previous money-losing incarnation—an all-female-hosted Mantovani ("easy listening") format. (During *that* period Metromedia had considered changing the station's call letters to WSHE.)

Meanwhile, similar transformations were taking place on the FM band in a number of major cities around the country. Some, specifically those in San Francisco and Los Angeles, were more genuine attempts at free-form, creative radio. Commercially speaking, however, WNEW was the best New York could come up with.

The new format was not an immediate financial success, but it grew and over the past few years has climbed to the top of the ratings. This, of course, means enormous advertising revenues (profits).

The only other such attempt made in New York FM radio was a weekly program on WCBS, hosted by a fellow who went under the name of "I. M. Flowers." Ironically, his real-life surname was Love. This venture didn't fare nearly as well as WNEW-FM's, however, and Flowers went the way of the "K"—to Jingle Village, no doubt.

WNEW, though more hype than hip, at least gave some amount of freedom to the person on the air. The music was good "progressive" rock, the disc jockeys' patter more human, if still basically banal. There were many fewer commercials, and the jingles themselves were less offensive and harsh. Occasionally the station even broadcast a public-service announcement that was really a public service. No doubt about it, WNEW-FM was one small step for the audience, one giant leap for the industry.

There's nothing that will get the creative juices of the media men flowing more quickly than a financial success. And so it wasn't long before another of the Big Boys came along to horn in on WNEW's territory. The American Broadcasting Company, home of *Eyewitness News*, Cousin Brucie, and *The Partridge Family*, decided that its FM outlets around the country should "go progressive." (I can imagine the network execs thinking that meant joining the labor movement.) As might have been expected, their idea of "going progressive" was a little peculiar: they would do mostly pretaped programing, standardized on all FM stations. The rest of the time would be turned back to the local station, for a little "local flavor." This format was a complete flop. The "kids" didn't listen, the advertisers didn't advertise, and the money didn't roll in, though a few heads rolled out.

So Allen B. Shaw, Jr., Vice-President of the ABC-owned FM stations and the brains behind the exciting new format, fearing for his head, made a legitimately bold move (though I suspect he was ignorant of its boldness at the time). He hired Larry Yurdin, the brains behind both the WFMU experiment and the highly publicized 1970 Alternate Media Conference at Goddard College, Vermont, as production director for ABC's O-and-Os (owned-and-operated stations).

Yurdin is the alternate media's number-one hustler: he sells his product like a Fuller Brush man, but he *knows* about radio, and his motives generally seem pure.* First of all, Yurdin convinced the parent corporation to apply to the FCC for a change of call letters, from WABC to WPLJ,** a change subsequently granted. The new call letters were intended to ease identification with the AM top-forty station as well as with the TV station that offers us *Let's Make a Deal.*

Taped programing was phased out and replaced by live personalities, including one or two from the old WFMU free-form days (now bona fide college grads). They hired a couple of good local newsmen, did away completely with offensive jingles, and began

---

* Yurdin is currently manager of Pacifica's Houston station, KPFT.
** Taken from the title of an old blues song, "White Port Lemon Juice," more recently rerecorded by Frank Zappa's Mothers of Invention.

broadcasting some of the boldest public-service spots ever heard on radio.

Playlists were a thing of the past at WPLJ, and Yurdin spoke both publicly and privately about the amazingly free hand Shaw was giving him. Even the people on the air were pleased, though some of them treated it all with disbelief, as though any moment the excrement would hit the ventilator. Even some of us at WBAI were unofficially sounded out about joining the "WPLJ team." All of us managed to resist the considerable financial temptation—mainly because we had a feeling it wouldn't last—but there was some cooperative programing between the two stations.

Now, despite the considerable skepticism of almost everybody, including the audience, WPLJ seemed to improve—that is, it became freer. When early in 1971 the FCC came down hard on "drug-oriented" song lyrics, no noticeable change in WPLJ's programing resulted. Then the station hired Alex Bennett (in real life, Bennett Schwartzman), who had just been fired by WMCA-AM amid a free-speech controversy. He was to host a morning talk show. (R. Peter Straus, the philanthropically liberal owner of WMCA and the other stations in the "Straus Group," maintained that Bennett's dismissal had been for "financial reasons." When Bennett—whose frequent on-the-air guests included Abbie Hoffman, Jerry Rubin, Paul Krassner, and Al Goldstein, editor of Screw—countered by offering to work for union scale, he received no reply from Straus. Bennett was replaced by a sports talk show.)

WPLJ began climbing, very slowly, up the rating scale. Everybody seemed happy. Now, finally, New York City had a commercial alternative—long overdue and far behind other major cities as it was.

Then Larry Yurdin left the station, swapping his WPLJ program director's job for an on-the-air spot at one of ABC's West Coast O-and-Os. There was no immediately noticeable change in WPLJ's format or freedom, but during the summer, the Federal Communications Commission, in a ruling involving a Des Moines, Iowa, free-form radio station, issued an edict stating that the "free-form" radio format "gives the announcer such control over the records to be played that it is inconsistent with the strict controls that the licensee must exercise to avoid questionable practices."

This was a very strong, if typically ambiguous government statement. Still, it does more than smack of repression. Now the FCC is a mighty powerful bunch, being so close there to Nixon and all. But still, as a group, no one is more powerful in this country than the media, especially the electronic broadcast media. The rise and fall of Spiro Agnew have made that clear. So did CBS, when its corporate apparatus was really at stake in the "Selling of the Pentagon," and it stood up to and beat not only the FCC but Congress and Nixon as well. The combined power of print and electronic media was victorious in *The New York Times* and the "Pentagon Papers" *vs.* the United States Government. By comparison, the Des Moines edict was fairly mild.

And so, if they cared to, the radio stations could have easily fended off this comparatively minor assault. *If they cared to.* But they didn't. Spokesmen from the parent corporations of New York City's two commercial "underground" giants made that abundantly clear in an article in *Broadcasting Magazine*, the industry's *status quo* organ, dated August 23, 1971:

*ABC and Metromedia, both of which operate several FM stations playing progressive-rock music, apparently are not troubled by the Commission's statement. By their own definition, their progressive stations are not, at the same time, "free-form."*

*Announcers at Metromedia's progressive stations have "enormous leeway" in what they play, said Willis Duff, manager of Metromedia's KSAN(FM) San Francisco. But at the same time, Mr. Duff added, "We have many checks and balances and he [the announcer] is subject to continuous review" of what he plays. Mr. Duff does not consider the Metromedia stations to be "free-form." "There are several stations in this country," he said, "calling themselves free-form or progressive that do not exercise adequate control. We do not fit into this category."*

But Metromedia's attitude was to have been expected. During the course of 1971 the station's programing went steadily down hill. It was fairly common knowledge that playlists, of sorts, were in use at WNEW, for after the FCC drug-lyric ruling, there was a glaring, if not complete, hole in its musical programing. The station's two best and most popular air personalities left—Roscoe to an early retirement on the French Riviera with his family and the fortunes he made narrating mayonnaise commercials for TV, and John Zacherly, probably the funniest and most underrated talent on radio today, to WPLJ, where it is rumored he has a "no management interference" clause in his contract.

As for the hipper-than-thou American Broadcasting Corporation:

*According to Allen B. Shaw, Jr., vice-president in charge of the ABC-owned stations, the network "is constantly tightening up" on its controls over the announcer. "We used to give them a lot more freedom," he said, but now "we are out of the free-form thing entirely." While the announcer is given the prerogative of selecting the order in which the records are played each hour, he said, each record scheduled for air play is predetermined by station officials. A disc jockey, he said, "may not be as scientific" as the station's music director in determining what records evoke the most popularity. Mr. Shaw added that "I agree in full with the FCC that free-form stations are not desirable"* [emphasis mine].

Shaw didn't mince his words, did he? And isn't it just great to have someone on hand who knows what too much freedom is? (ABC might do well to loan him out, for a fee, to the Nixon Administration.) I imagine it must have come as something of a surprise to the folks on the air at WPLJ that they were "out of the free-form thing entirely." I wonder what caused this complete turnabout in Shaw's attitude. A glance at WPLJ's share of the FM ratings for that quarter might provide a clue to Shaw's sudden loss of commitment.

And finally, folks, as I sat down that summer to begin writing this chapter, the phone rang. It was Vin Scelsa, WPLJ's afternoon man and a veteran of the WFMU experiment. "I've been fired," he said, as though delivering the long-expected news of a family tragedy. And he went on to tell me that he'd been called into the office late in the week of August 30 and informed of the new programing policy, which included our old friend, the playlist. He told me that on his first show the following week he announced that he would completely ignore the policy. It's hard to imagine why, but he was fired that afternoon.

I called WPLJ a couple of days later and spoke first with a producer who asked not to be identified. "It's not really a playlist," she told me. "They've just told us to throw in three or four cuts an hour from a list of top-selling albums. It's not really that restricting, especially for those of us who have come from commercial [radio] backgrounds. You have to remember Vinnie came from WFMU, where they had complete freedom. I guess he felt he just couldn't live with it. But PLJ's still the freest thing in town, except for BAI."

But a WPLJ interoffice memo, dated August 26, 1971, indicated a far more restrictive playlist. The memo, sent to "talent" from Mitchell Weiss, WPLJ's program director, addressed itself to the subject "Program Aids" (a euphemism worthy of the Pentagon):

*We currently have three record lists. There is an "A" list of hit singles. There is a "B" list of currently popular albums. Both the "A" and "B" lists appear on the first page of the copy book for your show and will be changed every Wednesday. There is a "C" list of standard progressive rock oldies also. This list at this time is in your hands and includes familiar tracks of Dylan, Beatles, Stones, etc. In the next few days we should have a "C" list available to you.*

*Basically we are asking you to use these program aids in the following way. Our revised Daily Music Playlist should be used for planning out each hour of your show. The key time zones—:00, :15 and :45 (indicated by the dot) should be filled with an A, B, or C track. All of the selections you play either in or not in the key zones should be marked in the box at the far right by an A, B, C, or an N for recommended tracks from new records. If the track is not in any of these categories, only then should the box be left empty. In the daytime there should be a minimum total of 4 B tracks. In our discussions I mentioned how these numbers vary at other times. In a four day period you should have played every item on the A list at least once.*

*We will be having weekly music meetings on Tuesday to discuss new records, recommended tracks, and the A, B, and C lists.*

*As the use of these aids develops, you should expect some refinements and alterations from time to time. For instance we are seriously contemplating* narrowing down *[emphasis mine] the B list as well as designing some kind of master chart for the B list to allow you to see at a glance the distribution of B record play prior to your show. . . .*

*Thanks for your cooperation.*

Later I spoke to Allen Shaw. He talked mostly about the declining ratings of not only WPLJ but also ABC's six other "progressive" FM stations. "We've been losing money right along," he said, "even when we were into that slick, ridiculous taped format." (Which is a hell of a thing for a man to call his own child.) "But in San Francisco in the eight months we've been free-form—that's three rating periods—we've gone from twenty-eight thousand [listeners] per quarter [hour] to six thousand in the ARB ratings. In *Pulse* we didn't show up at all [indicating a rating of less than six thousand listeners]. And in all the cities we're worse off than before."

Shaw blamed it on "the people." "Apparently there is a difference between what people say they want to hear and what they actually listen to. I'm mad at people being hypocritical with themselves and us. The problem isn't being free-form, or real, or controversial, but what the people will listen to."

Sound familiar? It should. It's the same argument, almost word for word, put forth for years by TV execs responding to criticism of TV programing—"We're giving them what they want." Remember?

But Shaw conveniently stuck his finger in the hole in his own argument: "Rating services, by their own admission," he said, "have difficulty reaching eighteen- to twenty-four-year-olds. They're too transient, always moving from place to place." And one of WPLJ's air personalities added, "Even if the rating service did reach one of our listeners, he'd probably hang up on them."

And so, when the FCC handed down its statement on free-form

radio it provided a nifty excuse, the little incentive that ABC brass needed to reinstitute the playlist. What comes next can only be determined by those in power within the media. They have the power to do something about an outmoded rating system, one that systematically excludes a large segment of the radio audience from the process that ultimately determines what they will have the chance to hear. While the FCC has the power to make statements and hand down rulings, the media have the power, if they will only use it, to fight back against those rulings which infringe upon their freedom, and the freedom of the public *to hear and see*. The Commission, following the same logic used in the drug-lyric ruling, could issue a statement banning lyrics that advocate love. After all, a 1971 survey showed that young girls are "sexually turned-on" by the lyrics of rock music. By banning those lyrics we could cut down on the problems of venereal disease, abortion, and the population explosion. Or, for that matter, any lyrics pertaining to peace, since they obviously give aid and comfort to the enemy and undermine the morale of our fighting men overseas.

How would the Allen Shaws, Willis Duffs, and Mitchell Weisses of the world react? If the latest ruling is any indication, you might do well to trade your radio for some more promising form of self-amusement. The industry has retreated from its small step ahead: the playlist has made its comeback. Stay tuned to this same station for offensive jingles and screaming disk jockeys—all brought to you on recording, of course.

## 11

# "Post Loses Post Post"

POST GETS POST POST was the facetious headline scrawled in grease pencil on the already amply graffitofied bathroom wall at WBAI in the spring of 1970, after I'd been offered the job of Director of Radio at C. W. Post College, a division of Long Island University. I had been with WBAI for nearly six years by then, and was seriously asking myself the question that every long-term WBAI person must sooner or later face: is it commitment to the institution that keeps me here, or fear of the outside?

I'd been given conflicting advice by those who had passed through a similar crisis. One twelve-year Pacifica veteran had, upon his departure a year earlier, warned me to "get out before it's too late." Another, who only weeks after I arrived left to work in television news, had assured me there was really no place to go but down; that one could make more money and reach a wider audience, but simply never again achieve the same level of freedom. Both pieces of advice made perfect sense to me then, and do now.

I thought I had already accomplished a good deal, if not all of what I could at WBAI, without yet having become bored and embittered. Though the administrative job at Post College was not precisely what I had envisioned for the future (Director of Radio at an obscure Long Island college was not exactly stardom), it was at least a chance to prove to myself that I was not serving a life sentence in the undisciplined, comfortable world of WBAI. The money wasn't bad either, by WBAI standards at least, and I would

get it on time—no small consideration after six years of promised paychecks. What was more, my schedule could be arranged so that I could continue my weekend show.

I'd originally been approached by a couple of students on the staff of the college radio station (WCWP-FM) after I had been on as a guest. There was a good deal of dissatisfaction among a large segment of the staff over the current orientation of the station. It had, for some years, been directed by an old radio professional and squash partner of the chairman of LIU's board of trustees, whose knowledge of modern communications seemed to end with the demise of the Dumont Network. WCWP, though holding a noncommercial, educational FCC license, was programed almost entirely as a commercial top-forty station, complete with its own set of station identification jingles, screaming disk jockeys, rip-and-read newscasts, and United States Army recruitment public-service announcements (PSAs).

C. W. Post College is located on a picture-postcard campus in Greenvale, Long Island, about thirty-five miles from Manhattan. Its student population is made up largely of the sons and daughters of upper-middle-class suburbanites who generally enroll in the School of Business. Post's tuition is among the highest in the state, its academic standards among the lowest. Walking about the campus in 1970, one got the feeling of having stepped back in time, about a decade, to the days of crew cuts, Corvettes, and penny-loafers. Post's most notable achievement during the activist years of the late 1960s and early '70s was its almost universal apathy while other students were striking against the Vietnam war and the Kent State killings. While other campuses were busily engaged in seeking greater student participation in the university decision-making process, Post's students were stuffing telephone booths.*

The school had been named after the late husband of Marjorie Merriweather Post, heir to the cereal fortune, but Mrs. Post, showing remarkably sound judgment, wanted nothing to do with her namesake.

A few of the more informed and active students, who had joined the radio-station staff a year or so before my arrival, were now the ones agitating for change. Already many other college radio stations both in New York and around the country had taken a new direction, greatly influenced, I believe, by WBAI and the other Pacifica stations. Their formats had begun to reflect the activism and creative energy of the times, both musically and in the spoken word, while WCWP was still emulating the insipid format of late 1950s and early 1960s radio.

These few students had, after considerable effort, awakened the administration to the need for change. They had submitted my name as a candidate for the position of director to the administration of the Theater Arts Department, under whose authority, for some obscure reason, the radio station fell. To my utter amazement, I was interviewed and offered thirteen thousand dollars a year for the less-than-full-time position. I suspect that my appointment was part of an attempt to update the institution's conservative image. My association with WBAI and the popularity of my program among college students provided the preferable image, while the relatively nonpolitical nature of *The Outside* did not seem to threaten the administration. The plans I articulated during the interviews, and those I was to carry out as director, were far from radical.

With some reluctance to leave WBAI, a bit of misplaced pride at having overcome my own lack of formal education, and a good

---

* Perhaps contributing to my later downfall were some facetious remarks I had made about the school on my WBAI program. Referring to the place as Howard Johnson University, I said that the year before, during the student strike, the campus radicals had given the administration a list of "sixteen nonnegotiable demands." "The administration turned them all down," I said, "and the students accepted that as a compromise."

deal of personal uncertainty about my ability to handle the job, I accepted and became Director of Radio on September 15, 1970. (Though I doubt that this date will long be remembered by students of the history of American communications, I note it here so the reader may, in the end, calculate the brevity of my academic career.)

The setting was ideal. The radio station, housed in its own building, had large and well-equipped studios and production facilities, far superior to those of WBAI at the time. I was provided with an impressive (to me) office, and an extremely capable and dedicated secretary/administrative assistant, who remained devoted until the end and beyond. The budget, though small, was adequate for that year, and I was led to believe more money would become available upon submission of my revised budget plan. Dr. Julian Mates, a career academician and newly appointed Dean of Humanities, a nervous, tweedy, pipe-smoking, absent-minded-professor type, had, in our preliminary discussions, assured me autonomy and academic freedom, while acknowledging his conservatism and our generally differing philosophies. Mates was a gentle, sincere soul, though it is questionable whether academic freedom took precedence when his head was on the chopping block.

My immediate superior was Dr. David Scanlan, who was beginning his first semester at Post as director of the Theater Arts Department. He was a bespectacled, mild-mannered, quietly trembling fellow about forty years old, and his background consisted primarily of directing small-town theater groups, plus some teaching. Scanlan knew nothing about radio, and cared even less. Both he and Mates willingly admitted that, to them, the radio station was simply a headache which was constantly torn by political strife and which somehow, quite accidently, had fallen into their laps.

Indeed, WCWP was treated like, and was in fact, an afterthought. Housed in the Benjamin Abrams Communications Center building, it was the gift of said Benjamin Abrams, whose background and connection with the college are obscure, except that he was wealthy and probably motivated by the tax-deductibility of such a gift, as much as by his dedication to the fine art of communication. (At the time of its inception, around 1965, there were still many unassigned FM frequencies which the FCC readily made available to almost any financially secure nonprofit institution, especially an educational one. Today there are no longer *any* FM frequencies available in the New York area, and precious few elsewhere in the United States.)

The station's first director had been a faculty member in the Theater Department and his principal qualifications were his interest and availability. Though he had no radio experience, he was, even then, an enthusiastic listener and supporter of WBAI. (Had the administration known about this, he undoubtedly would have been disqualified at the start.) Had he been permitted to translate his theories into practice, WCWP might have become a pioneer in college radio. But at first hint of his unorthodox approach (the institution of a late-night jazz program), he was relieved of his radio responsibilities with a pat on the head and a push back to the front of the classroom, where he remains to this day. His eventual replacement was a friend of Post's Board Chairman, John McGrath, who also worked as a part-time weekend staff announcer at a commercial New York City FM station. Though I replaced him, he retained the title "Director of Broadcasting," plus his fifteen thousand dollar annual salary, apparently because of his personal relationship with McGrath, while his responsibilities became those mainly of a part-time instructor in "communications." (In fact the entire "communications" faculty consisted of three part-time instructors, all radio traditionalists whose backgrounds were mainly in minor positions with small, local Long Island commercial stations.)

My appointment was greeted with astonishment by those on campus who knew something of my background and of the history of both WCWP and the college in general. (Two instructors very highly regarded by the students had recently been denied tenure on dubious technical grounds. Both, coincidentally, had been outspoken supporters of a token student strike that had taken place the previous spring.)

My initial act as WCWP's director was to call a meeting of the entire student staff (attended as well by Mates and Scanlan) at which I assured all "factions" that I intended to run a station that was closed to no one, and at which all kinds of radio could be learned and practiced. Though I intended to make major changes in the mostly top-forty format, all those who currently had programs would continue. In order to open time in the schedule for new programing, all daily shows would be cut back to one to three times weekly. Additionally, programs now held by those who were no longer students would be canceled in order to achieve greater student participation.

The logical first step in building a relevant college station, it seemed to me, was to involve the rest of the campus—to draw from its resources and serve its needs. As it was, the radio station stood isolated from the rest of the college, a world unto itself (as radio stations tend to become), serving only the handful of students constituting the staff. Few students, faculty, or members of the local community listened to it, and there was absolutely no participation in programing on the part of students or faculty in other departments.

We devised a format that encouraged students and faculty from other departments to participate as producers of biweekly "magazines" of the air. In the course of recruiting department members —music, theater, English, sociology, journalism, and others—we encountered a good deal of resistance and suspicion. Many had made earlier attempts at participation, and all had been turned away, either directly or indirectly by lack of cooperation from WCWP's student staff, which had the apparent sanction of the former director. These creative souls remained frustrated by the experience and were bitter toward the station.

But we were able to persuade some that there would now be genuine change, and by November 1, 1970 (the target date for the beginning of the new format), we had involved nearly one hundred students and faculty in new programing. Additionally, we created a fifteen-minute daily "Opinion" slot, open to all shades of campus and community attitudes, on issues ranging from the quality of college lunches to terrorist bombings, to local and national electoral politics.

The news department was completely altered, from five-minute, on-the-hour rip-and-read newscasts, to half- and quarter-hour news-casts, gathered from numerous sources, written and reported by each individual newsperson. The news programs were heavily oriented toward local and campus news, and we purchased several portable cassette tape machines for the reporters' use in covering local events. I even proposed adding a Reuters news wire, to aug-ment the already existing AP and UPI wires, which would serve to expand and offer a different approach to international news. Our purpose in making these changes was not simply to create a comprehensive news program, but to begin to educate a generation of electronic journalists—not merely well-trained voices, as is mostly the case in broadcast journalism today. We looked for reporters and writers, not just vocal cords.

By November 1, six weeks after my appointment, we had issued the "Winter Program Guide," and the new format was under way. The "Official Page" of the guide listed the names of the student staff members, fifty-seven of them—nearly twice as many as the previous year—and contained the following optimistic message to our listeners from the student programing department:

. . . the staff and management of WCWP has been imbued with a new pride and sensibility. . . . We now find that an educational radio station does not necessarily have to be a boring one. . . .

You will find WCWP a mixed bag. The concepts of continuity and repetition have given way to those of creativity and diversifica-tion. As the broadcast day progresses, the programing does the same. The greatest change in programing is the implementation of a weeknight "magazine" format. Here the listener will find every-thing from Strauss to Seale. These magazines . . . are designed to entertain as well as inform.

Some examples of the magazine programs from the WCWP winter programing schedule:

SYNTHESIS: AN EFFETE REVIEW—The bi-weekly magazine of the Arts. Included are: radio drama productions, tape montages, poetry, live theater drama, and discussions of plays, music, and films.

GIRYLS, GIRYLS, GIRYLS—A program produced by and for women.

FOR WHAT IT'S WORTH—produced by members of the Sociology Dept. An intensive look at some contemporary American problems.

POST SCRIPT 70's—Devoted to pursuing all aspects of the Humanities. In our exploration we will employ music, live discussion, tapes, dramatic presentations and any other method in order to create conflict and/or resolution in the areas of our investigation.

THIS MOMENT—Music from 16th-century lute dances to John Fahey; Pat Sky to the Incredible String Band.

DONA NOBIS PACEM—Live classical music performed on campus.

SATIRICON—OR A TREE GROWS IN BROOKVILLE—. . . some of the sickest satire east of Secaucus.

In addition, the schedule included regular jazz, top-forty, and free-form programs, live coverage of campus sports events, and, when possible, concerts and live guest lectures. With the dedicated help of a handful of students, the accomplishments of those first hectic six weeks exceeded my expectations.

Our enthusiasm was not shared by all. I was unable to redirect the considerable energies of the top-forty people (as they came to be known, conjuring up rather bizarre images) into the new format. Their initial insecurities remained strong, and what cooperation they extended to the new producers was given grudgingly. Some of the new producers even claimed to have had their productions sabotaged, and while at the time I treated their accusations with skepticism, by the time I left the peaceful halls of academia no tales of intrigue seemed too far-fetched. Despite my stern warnings and constant reassurances, many of the student staff continued to work against us, and considerable time was consumed by internal politics. In contrast, WBAI looked like the Peaceable Kingdom.

My contact with Mates and Scanlan during these first weeks was minimal, limited mainly to telephone calls regarding administrative details. They seemed only too happy to be relieved of the burden of WCWP, which had seemed, like an albatross, to hang from the neck of the administration. Though neither Mates nor Scanlan ever inquired, I frequently volunteered reports on the progress of the station, reports that seemed to please them. Copies of all important policy revisions, such as those regarding public-service announcements and the use of "frank language," were passed along to them for their approval. Their silence on these matters, I assumed, constituted that approval.

By the third week in November—three weeks into the new programing format—all appeared to be going well, save for the continuous internal bickering. Though my secretary and a few of the staff had warned me that a number of the students were still determined to "get rid of me," I chose to ignore the warnings, which I viewed as paranoiac, and to concentrate my efforts on programing development. I dealt with each internal conflict as it arose, choosing—naïvely, it appears in retrospect—to shrug off the talk of conspiratorial plotting. I had been hired to develop an educational radio station, and I believed our programing spoke eloquently to that end. Both Mates and Scanlan, in no uncertain terms, spoke with approval of what they heard.

On Wednesday prior to the four-day Thanksgiving holiday, Dr. Scanlan came to my office. Again he enunciated his satisfaction with programing and the over-all direction of WCWP. (Scanlan, despite my lack of official faculty status, had earlier appointed me to the Theater Arts curriculum committee to help formulate courses in communications for the following semester. Though I had been asked in our initial negotiations to teach several courses, I had, on the grounds of sheer terror, turned down the offer. Had I, at the time, recognized the generally rock-bottom level of too many of those assigned the title and the task of "educators," I would, no doubt, have felt more adequate.) He had stopped by the office, he said, to discuss some general complaints about a "lack of administrative control." When I asked him to translate the euphemism, "lack of administrative control," and to reveal the source and nature of the complaints, he was unable, or unwilling, to do so. (I might mention, however, in Scanlan's semi-defense, that his mumblings only occasionally reached comprehension, which I found particularly curious in view of his theatrical background. Sometimes it seemed as though his speech pattern was merely an esoteric code.)

At any rate, I rearticulated my belief in minimal administration, assuring him that while I did not believe in anarchy, I did hope to provide an open atmosphere in which the students could develop creatively along whatever lines they chose. Further, I said, there was ample room within the framework of the station for differences both political and creative, and whatever conflicts existed could be worked out internally among the students, so long as they did not interfere with the work of others and the station's programing in general. I added, however, that if he could provide me with details, I would look into the specific problems. I had as yet removed no one from the staff, but I recognized that it might become necessary to do so.

Scanlan seemed tranquilized and satisfied by our conversation and by my assurances that I felt all was under control, if not calm. Although calm would more easily have been attained than creativity, I foolishly believed the administration to be more concerned with the latter. Had I been less naïve, I doubt that I would have appreciably altered my approach. I left for my Thanksgiving holiday with no more than the usual amount of anxiety.

Early Monday morning, November 30, Scanlan phoned me at home. In a voice betraying no more uneasiness than was normal for him (he always spoke and looked a bit as if he expected, at any moment, a large object to fall from the sky and strike him while his back was turned), he asked me to stop by his office before I arrived at the station that day. I can no longer recall my precise speculations over the next couple of hours, but I do believe I rolled over and went back to sleep.

I was being dismissed, effective immediately, he said, for "irresponsibility" and "placing the station's license in jeopardy." Once again Scanlan could not, or would not, supply details. Though stunned, I decided initially against waging a battle; I would collect the three months' severance pay called for in my contract, and take a long-awaited vacation.

Scanlan understandably looked a bit more awkward and uneasy than usual, suspecting, perhaps, that I would be the large object to pounce on him. Oddly enough, I felt sorry for him, having on several occasions been in a similar position in my role as WBAI's chief announcer. I tried to reassure him, stating that I planned to mount no campaign for reinstatement. Then I asked him about my severance pay. "You'll get two weeks," he said. I told him I was sure my contract called for three months. He replied that I was being given only two weeks because I had placed the station's license in jeopardy. Its truth aside, not only was that irrevelant, it was also a legal breach of contract, which I hastily pointed out to him. He backed off rather abruptly, saying he would see to it that I received the full amount due.

As satisfied as I could be under the circumstances, I strolled across campus to the Benjamin Abrams Communications Center, decided I would call a hasty staff meeting to announce my dismissal, clean out my desk, and head home—in time, I hoped, to stretch out in bed and enjoy the luxury of being there for the six-o'clock news. (Ordinarily, I arrived at the school at about one P.M. and left some time after ten in the evening.)

It seemed clear that the decision to fire me had not been Scanlan's; he had simply been appointed the executioner. Surely it was a decision he could not have come to in the few short days since our last meeting. Had he felt the station to be in such irresponsible hands, I would have received the appropriate warnings much earlier, if only through the grapevine. The same would have been true of Mates, from whom I had received not the slightest indication of concern. I believed (with some substantial later evidence to back this belief) that the decision had been made, under pressure from members of the Board of Trustees, by Robert Payton, President of C. W. Post College and a former minor political appointee under the Kennedy and Johnson Administrations. (Payton had been United States Ambassador to the Cameroons, which, until my enlightenment, I had believed to be a Passover cookie. He had a reputation for walking casually about campus with his jacket flung across his shoulder, looking like the perennial candidate filming a sixty-second TV spot.) His reputation for decision-making based on political expediency was widespread on campus. He had seen to it that mild-mannered Professor Mates would take the responsibility for my dismissal—and related events—perhaps as the rather nominal debt to be paid by Mates in return for Payton's earlier appointment of him as Dean of the Humanities Division.

I arrived at the station and announced my dismissal, to the astonishment of most of the staff, and the quiet joy of others. I told them I planned no action and suggested they go about the business of running the station, which was scheduled to begin its broadcast day at four P.M. I returned to my office, noting that the plastic plaque machine-engraved with my name, which had been above the one which said DIRECTOR OF RADIO, was missing from my office door.

Shortly after four P.M., several of the staff burst into my office in a state of panic; they were unable to turn on the transmitter to begin broadcasting. I phoned Scanlan, who informed me that, as a precaution, all power to the transmitter had been cut the previous night. He did not make clear what the action had been a precaution *against*.

The students, furious, demanded an immediate explanation. A few minutes later, armed with the following memoranda, Dr. Scanlan arrived at the station. The first, addressed to the staff of WCWP, was from Julian Mates, and stated:

The accompanying notice outlines the reasons for an immediate (Nov. 30) suspension of broadcasting. I know how hard most of you have worked at the station, how much of your lives you have devoted to it. My decision was taken after consultation with many people, but it is no easier for that. I hope that most of you will be back with the resumption of broadcasting. For a fuller discussion of the factors that went into my decision, may I invite you to a meeting tonight (Mon.) at 6:30 P.M., in Humanities 100.

The other notice, addressed to the "C. W. Post Community," helped little to clarify things:

An intensive study of the radio station at C. W. Post has revealed serious flaws, largely brought about through the lack of both guidance and guidelines—a polite way of saying I've goofed. With the Fall semester drawing to a close, and in order to protect our station license (not to mention our reputation), I have thought it best to suspend all broadcasting while a Board of Governors is established to help determine policy. The station has been an arm of the Theater Arts Department; in future, it must serve the entire Center, and the governing body will help set the boundaries for policy, for programing, for participation. The Board will be made up of faculty and students from all areas of the C. W. Post Center, and the Director of Broadcasting will report to it. The level of technical competence should be improved as additional courses in Radio are added to the curriculum, courses which will feed well-trained students to the station. A target date for resumption of broadcasting is January 1.

The students, well-attuned to administration newspeak, were not a bit satisfied. Scanlan said that Mates would explain further at his proposed meeting that evening. The students replied that they would not attend, fearing they would be locked out of the station. Scanlan summoned Mates to WCWP's main studio for a meeting with the students. I remained in my office with a couple of sympathetic faculty members who were urging me to stick around and fight back. Meanwhile, unbeknownst to any of us, campus maintenance people were—at the direct order of Mates—changing the locks on all doors leading into the radio station.

The student staff emerged from the meeting an hour later, reporting that neither Scanlan nor Mates had presented any evidence to substantiate their charges. They became even further incensed when they realized the locks had been changed, and concluded that the meeting had been called primarily to divert their attention while the maintenance crew made certain they would not have access to the building again. Apparently there was some merit to their earlier belief that the "Humanities 100" meeting was merely a deceptive device to lock them out. The majority of the student staff voted to remain in the building until the station was returned to the air and I was reinstated as its director.

Meanwhile, back in my office, my own rage was growing with each new insight into the actions of the administration. It seemed to me, no matter what the true nature of their beef, that my own dismissal was sufficient. What motivation, other than simply silencing the students and preventing the public disclosure of the incident, would there have been for cutting the station's power in the middle of the night? How much more blatant and obvious could they make their denial of free speech? There was simply no justification for denying these students access to a radio station which they had built.

All afternoon sympathetic and outraged faculty and students paraded in and out of my office. Late in the afternoon I issued what I then thought would be my first and last statement:

Ironically, I came to C. W. Post College because I believed that working with a radio station in an academic setting would be an opportunity to develop creative and innovative uses of the radio medium. To my disappointment, I found instead an atmosphere that was not only closed to new ideas but which also seems to be controlled by those with political and financial power.

If, as has been charged, the station's license and reputation have been jeopardized under my directorship,* this should have been brought to my attention prior to my dismissal. In my limited telephone conversations with Dean Mates and in my one discussion with Dr. Scanlan in my office (on Nov. 25, 1970), they brought to my attention administrative questions of certain broadcasts. I agreed in principle with these criticisms and said that I would try to find solutions to these problems. They never indicated to me in any way that these questions were serious enough to warrant my dismissal.

In addition, if, as Dean Mates alleges, our license and reputation were in jeopardy, it seems odd that I never received any indication of this either from the Federal Communications Commission or from listeners. Dr. Mates has stated that he received several letters of complaint. Why, then, have these never been brought to my attention? If they were genuine, spontaneous letters of criticism from our listening audience, it seems strange that they were addressed to the Dean of Fine Arts and not to the radio station. I believe, there-

* Memo to C. W. Post Community from Dean Julian Mates, subject, WCWP, Nov. 30, 1970.

fore, that these letters were the result of an organized campaign directed against the present direction of the radio station.

The basic question here is: does the University want a free and open radio station or does it want a radio station that serves as the arm of the Public Relations Department of the University? The actions of the administration would indicate the latter.

I believe that the interests of the "University" are at odds in this situation with the interests of its students.

My own determination to remain outside of the political struggle was waning. In large part, I was encouraged by Patti Baltimore, a sympathetic faculty member who herself had recently been denied tenure for what appeared overtly political reasons, and who had invited me to stay at her home nearby. And on my own I realized that the actions of the administration could not be justified. The following day, December 1, I returned to the campus in the afternoon and addressed a hastily planned rally of several hundred students on the lawn outside of the Humanities hall.

Later that day Dean Mates, under heavy criticism from both campus and local press (*Newsday* and the *Long Island Press* carried front-page stories on the controversy, and *The New York Times* and *Daily News* carried items in their radio sections) for his inability or unwillingness to back his charges with specifics, issued yet another statement addressed to the "C. W. Post Community":

I have been asked to give, in some detail, my reasons for the temporary closing of the College's radio station. The F.C.C. "can revoke or suspend a license if programing, in their opinion, has not been for the public good" [a not entirely accurate quote]. Too, I am informed that any broadcaster who plays obscene material on the air can lose his own license for so doing. In my opinion, several programs in recent weeks have been obscene and without any redeeming social or educational value. This past weekend pornographic literature was read over the air; a week or so ago a two or three-minute segment, at 7:00 P.M., was aired featuring a woman having an orgasm [emphasis mine]. There have been other instances in my judgment and on the basis of phone calls and letters I have received we were in imminent danger of losing our license. At the moment, there is a long list of applicants for an FM license. If we were to lose ours, many years might pass before another license might be obtained.

Furthermore, there are serious doubts as to whether the station is fulfilling its educational function. There is also confusion as to what that function is. Students at WCWP have had no guidelines. They are accountable to no one. What I want is a governing board made up of faculty and students who will establish such policy.

In a *Village Voice* piece (page three of the December 10, 1970, issue) entitled "Post Loses Post Post," reporter Mary Breasted described the tapes—both of which, along with some of my files and my complete set of keys, had mysteriously disappeared from the station—Mates considered "obscene and without any redeeming social or education value":

Somewhere in the administrative offices of C. W. Post College, or somewhere among the possessions of a friend of the administration of the college, lies a two-minute radio tape of a girl breathing heavily and with progressive urgency followed by a voice explaining, "You've just been listening to the sounds of the 100-yard dash."

"That, much to my surprise, the administration recognized as an orgasm," C. W. Post's recently fired radio station director will wryly tell you by way of explaining why the tape disappeared mysteriously from the studio last week.

. . . One of the nuttier products of the students' newly creative efforts was the simulated orgasm tape. A finer and much more difficult creation was an hour-and-a-half tape satirizing sexual attitudes and roles in America. Put together by three students for a weekly satire show (a Post innovation) called "Satyricon or A Tree Grows in Brookville," this tape, broadcast November 27, started with selected readings from "The Perfumed Garden," an Arabian book of erotic wisdom which any reader of the Grove Press paperbacks knows to be pretty tame stuff. Nevertheless, the "Perfumed Garden" selections were the spiciest things on the tape, vivid descriptions of the ideal male and female. The rest of the tape contained readings of one student's morbid teenage love poetry,

mock letters and answers from and to the lovelorn, a vocal enact-
ment of a romance comic story, and a mock FCC bust of the station. All interspersed with 1950's bee-bop music.

The whole thing was quite funny. It moved along well, and the
students who played the various characters sounded as though they
really relished their parts. At various intervals they warned their
listening audience that some segments of the program might be
considered offensive, thus giving the prudish a chance to tune out.

Regarding Mates's claim that the station was "in imminent danger of losing our license," a December 3, 1970, piece in Long Island's *Newsday* said, "Spokesmen for the Commission said Tuesday
that no complaints had been received and that no action was con-
templated."

In addition, to my knowledge (and most certainly I would have
known), the station had received not a single letter of complaint
or criticism while under by administration. And, again according
to the *Voice* piece, "Mel Rosen, the station's public relations
manager, said there had been many letters of complaint during the
previous manager's time. 'There were numerous complaints against
the Top 40 shows,' he said, 'and that the station wasn't doing its
job.'"

Additional inquiries made to the Federal Communications Commission revealed that they had received no complaints about
WCWP's programing in over a year. (The Commission maintains
complete files on every radio and television station throughout the
country. As standard policy, they forward copies of all complaints
to the station managers, along with a request for explanations of
any broadcast in question. These letters, along with the explanation,
are then placed in the station's file for review at license-renewal
time.)

The students, now calling themselves "the collective," continued
their sit-in at the Benjamin Abrams Communications Center,
though regular classes continued in the building, and free access
to the station was denied to no one having business within. A
twenty-four-hour-a-day "security guard" was ordered assigned to

the building, and a memorandum dated December 1, from the college's chief of security, stated: "The building will remain open twenty-four hours a day. . . . Security will not attempt to evict any Radio Station personnel from the building, unless damage to the property occurs."

On December 2, apparently realizing that their action would not simply blow over, President Payton appointed a "special student-faculty committee of inquiry into the operations of WCWP." The appointed committee consisted of six faculty members and six students. One of the six students on this "impartial" committee just happened, by chance (out of a student population of nearly ten thousand), to be Francis L. McGrath, son of C. W. Post's attorney, and nephew of John McGrath, chairman of C. W. Post's Board of Trustees.

On December 3, the campus chapter of the usually cautious American Association of University Professors met and unanimously passed a resolution demanding my immediate reinstatement, pending a complete AAUP investigation, and rejecting Payton's appointed committee on the grounds that: "1) It is at variance with the faculty council resolution that specifies the faculty members must be elected; 2) One faculty member is a department chairman; and 3) It is unrepresentative of the general C. W. Post academic community."

Later that week, by a 24–15 vote, an underattended meeting of the faculty passed a resolution which condemned the "unilateral action of the administration in terminating Mr. Steven Post's employment at C. W. Post Center. We believe that his academic freedom has been breached." It further called for my immediate reinstatement, pending the AAUP investigation.

But the administration showed the same disregard for the opinions of the faculty as they had for those of the students. It was becoming clear that whatever powers had truly been behind the shutdown and my dismissal would be swayed by neither justice nor public opinion. The issue was receiving wide coverage on many local and New York City radio stations, as well as continuing coverage in the newspapers, especially *Newsday* and the *Long Island Press*. The *Press*, in fact, had done a bit of investigative reporting and, in a piece dated December 3, 1970, seemed to confirm my suspicion that Mates had been pressured into firing me:

*The Press, meanwhile, has learned that a group of former students, who until recently were involved in high positions of station operations, met with college president Robert Payton to complain about the way the station was being operated.*

*One of those involved charged that Steve Post, director of broadcasting, whose firing accompanied the shutdown, had usurped power from an executive committee of students which had a voice in station policy. Another former student got involved after a show he had been doing at the station for five years was canceled as Post instituted programing changes.*

*Payton met with station staffers after the two former students approached him and, according to a station spokesman, the entire staff said they liked the way Post, who describes himself as "a lovable dictator," was directing station operations.*

*One of the two former students, it was learned, was believed to be interested in the position to which Post was appointed.*

*Dr. Julian Mates, dean of the humanities division and the official who ordered the station shut down on Monday, said the two people in question had not met with him, nor had he spoken with either in the past three or four weeks. He said, however, he was surprised recently to hear on the air one of the former students believed involved, "because I didn't think he was a student any longer."*

And, in the same newspaper the following day, December 4:

*At the Brookville campus faculty meeting, C. W. Post president Robert Payton told the meeting the decision to shut down the station and fire Post was strictly Dr. Julian Mates's. But Payton added, "I must say I agreed with it."*

Meanwhile, as the committee met in closed session and the supporting faculty applied what small pressure it could, or would, exert, the protesting students laid plans for a massive rally to be held on campus the following Monday, December 7. The level of energy and the sense of mission displayed by the collective during these days and the weeks to follow was extraordinary. The sit-in continued twenty-four hours a day: at night bodies were sprawled about on the floor, draped over chairs in the studios and my office, catching a couple of hours' sleep here and there before setting about the next day's tasks.

On December 3, one of the students phoned attorney William Kunstler to see if he would be available to speak at the Monday rally. Later we spoke, and he told me he'd been following the situation in the media (we had met briefly on several occasions when he had appeared on WBAI). He said that he didn't know whether he could make it to the rally, but that, according to what he'd read, he felt there was potential for a legal case, based on the violation of the rights of the student staff, faculty, and students of C. W. Post, and those of the listeners to the station under the First and Fourteenth Amendments to the Constitution. Kunstler spent more than an hour on the phone with me and a number of the students, and said he would prepare and send off to me a set of legal briefs the following day. Which he did.

The following day the collective organized a press conference for the evening (December 3), to announce Kunstler's possible participation.

WCWP's enormous main studio that night was packed to capacity with media people, most of the campus's fourteen-man Public Relations Department, faculty, and representatives of the administration. As I announced the impending legal action under the direction of Kunstler, one could see physical evidence of astonishment on the faces of the administrators and public-relations people.

The following day we received the legal papers which named as defendants Payton, Mates, Scanlan, and the entire Board of Trustees of Long Island University. The brief asked for "temporary injunctive relief" (the immediate resumption of broadcasting), pending a hearing on the merits of the case. The suit asked for a declarative judgment against the defendants.

For several days we held off on a decision to file the court action, hoping the student-faculty committee would complete its investigation and issue a report. But it was taking its sweet time. By Monday, the day of the rally, we had decided to go ahead with the case.

The rally itself was one of the largest in the history of C. W.

Post, attracting almost three thousand participants (though media accounts varied in reporting the number of people present). Kunstler, after all, did show up. On Tuesday, December 8, *Newsday* reported on the rally:

*"Shutting down a radio station is not a hell of a lot different than burning a book," Kunstler told a rally . . . yesterday in the campus auditorium. He said that perhaps college officials acted in good faith in closing down the station, but that "there always is a reasonable excuse for every excess of tyranny."*

*. . . Kunstler said he was interested in the case because he believed it involved "a gross violation of the freedom of speech provisions of the First Amendment and was one of the latest in a series of attacks on student populations." He said that "all around the country student newspapers and radio stations are beset by administrators."*

*Other speakers included Hedda Garza, the Socialist Workers candidate for Nassau County executive last month; Alex Bennett, WMCA announcer; Pete Fornatelle, WNEW-FM announcer; Paul Krassner, editor of the Realist magazine; Kent Mostin, the station's morning director; and Post. The rally ended with David Peel and the Lower East Side, a rock group, singing several choruses of an improvised "C. W. Post Blues."*

The Board of Trustees prepared itself for the legal battle by retaining as defense counsel Emile Zola Berman, whose name had most recently been connected with defense of Sirhan Sirhan, the convicted assassin of Robert F. Kennedy.

On the morning of Wednesday, December 9, we filed suit in Brooklyn Federal Court, while several dozen supporters picketed outside. It was our unfortunate luck to have the case added to the docket of Judge John R. Bartels, considered to be the Brooklyn court's most conservative justice. (When we entered the building Kunstler said something to the effect of "God help us if we get Bartels!" God did not help us. The court clerk who formally presented the papers to Bartels in his chambers reconstructed for us an approximation of the judge's words when he saw Kunstler's signature on the memorandum of law: "I'll have none of his she-nanigans in my courtroom!")

As far as we could determine, it was the first legal case in which listeners had acted as a class, claiming a violation of their right to "hear" under the First Amendment. (Three listeners not connected with the college, who had been closely following the development of WCWP, acted as plaintiff-representatives of the listening audi-ence. Indeed, for all I knew, they might have *been* the listening audience.) The order to "show cause" was returnable on Decem-ber 18.

The press continued to give the story broad coverage. I suspect this was due in large part to Kunstler's involvement and the fact that Post's Board of Trustees had retained an equally big legal name to act on behalf of the defense. Berman's reputation as a "judge's lawyer" was to serve him well, while Kunstler's reputation as a "disruptive radical," increased enormously by the recently com-pleted case of the "Chicago Seven," did not help our chances. In the end the legal "glamour" of the two men served to obscure the real issues. But for now they focused press and public attention on the case, a circumstance which deeply distressed the college admin-istration, and Payton in particular.

Kunstler, of course, could do nothing about the reputation he had gained at least partially through press coverage of the Chicago trial. But despite the pressure of what must have been dozens of other extremely important political cases he was handling at the time, all without fee, Kunstler's legal papers appeared always to be well researched and in order. (Often we would hold hasty meetings at airports or railroad stations, or running across town, or as Kunstler drove recklessly at death-defying speeds in his red Beetle, pounding on the horn in traffic, shouting impatiently at anyone or anything impeding his forward progress. He could not bear to stand still for a moment.)

Some excerpts from Kunstler's "Memorandum of Law on Be-half of the Plaintiffs," submitted that day to Judge Bartels:

*In 1969, the Court for the first time fully considered the First Amendment issues involved in radio broadcasting. Red Lion Broad-*

casting Co., Inc., v. F.C.C., marked an important advance in First Amendment theory concerned with affirmative promotion of the system of Freedom of expression. In particular, Mr. Justice White, speaking for a unanimous court, emphasized that:

> This is not to say that the First Amendment is irrelevant to public broadcasting. On the contrary, it has a major role to play as the Congress itself recognized . . . which forbids FCC interference with "the right of free speech by means of radio communication." Because of the scarcity of radio frequencies, the Government is permitted to put restraints on licenses in favor of others whose views should be expressed on this unique medium. But the people as a whole retain their interest in free speech by radio and their collective right to have the medium function consistently with the ends and purposes of the First Amendment. It is the right of the viewers and listeners, not the right of the broadcasters, which is paramount.
> . . . It is the right of the public to receive suitable access to social, political, esthetic, moral, and other ideas and experiences which is crucial here. That right may not constitutionally be abridged either by Congress or by the FCC.

A fortiori, it follows that "that right may not constitutionally be abridged" by licensees or other private persons.

The First Amendment protection of free speech has been held in other cases to extend specifically to listeners. Lamont v. Postmaster General. On April 7, 1969, the Supreme Court emphasized in the case of Stanley v. Georgia, that:

> It is now well established that the Constitution protects the right to receive information and ideas. This freedom of speech and press . . . necessarily protects the right to receive . . . (citing cases). This right to receive information and ideas, regardless of their social worth, see Winters v. New York, is fundamental to our free society. . . .

Meanwhile, back at the campus, the investigating committee continued to investigate. Refusing to submit to pressures to issue its findings early enough to head off legal action, the committee issued a statement on December 7 which said that "substantial progress has been made" and stated that it expected to issue its report to the president on Wednesday, December 9.

On Kunstler's advice, we went ahead with the case, with the knowledge that the suit could be withdrawn at any time if the committee's findings were acceptable and the administration agreed to their implementation.

As promised, on December 9, 1970, Payton's appointed student-faculty investigating committee issued its report. The findings, almost completely in our favor, were perhaps more astonishing to me and the collective than they were to President Payton. But no one, as we were to find out before long, was more overwhelmed than the Board of Trustees. The report of the Committee read, in part:

*It is the opinion of the undersigned members of the special student-faculty committee of inquiry that:*

*1. Although the material which was presented to the committee did not fall within the personal definition of obscenity of the undersigned committee members, it is our opinion that at least one of the broadcasts contained material that was in poor taste.*

*2. Based on the information received by the committee, no governmental investigation was being conducted with regard to WCWP during Mr. Post's employment. . . .*

*5. Mr. Post instituted and encouraged policies whereby a full range of opinion had the opportunity to be broadcast on WCWP. These policies, moreover, appeared to encourage creativity and versatility of programing.*

*6. In our opinion, the administration has the right to terminate letter of appointment, but the action taken in this instance seems to have been precipitous. . . .*

*8. Although Mr. Post might have acted more swiftly and more decisively in dealing with organizational problems encountered at WCWP, he did not have adequate time fully to cope with these problems.*

Further, the committee recommended that a broadcast panel made up of three students, three faculty members, and the Director of Radio be elected to set broad guidelines for the broadcasting of WCWP; each member would have a single vote on the panel. It recommended that in the interim, until such a panel could be set up, a committee composed of the Director of Radio, the President of the Student Government Association, the Chairman of the Faculty Council, and the Vice-President of Student Affairs determine policy for the station. And finally: "It is the recommendation of the undersigned that Mr. Steven Post should be rehired as Director of Radio under the conditions of employment recommended above."

There were three dissenting votes, including Francis L. McGrath.

The following day *Long Island Press* and *Newsday* carried front-page stories entitled, respectively: REPORT PAVES WAY FOR COLLEGE RADIO STATION TO REOPEN and POST IS URGED TO REHIRE BROADCASTER, while the *New York Post* erroneously headlined its report, C. W. POST FM STATION IS ON AGAIN, and reported, with the requisite number of typographical errors, that the C. W. Post College radio station had already returned to the air, and that President Payton had agreed to implement the recommendations of the committee. Even *The New York Times*, bastion of accuracy, jumped the gun in a story headlined, COLLEGE DECIDES TO REOPEN WCWP. (The *Times'* coverage, in fact, had been generally shoddy and inaccurate.)

The reaction of most of the collective was joyous, though some, who had warned all along of the futility of fighting the college power structure, remained prophets of doom.

The next several days, including the weekend, were spent in meetings with Payton and members of the "temporary broadcast panel" to hammer out "details for implementation" of the committee's recommendation. (By this stage I not only understood their bizarre language, but was becoming rather fluent in it.) It was clear that Payton would attempt to interpret those recommendations in such a way as to limit my authority and the students' freedom to operate WCWP. The sessions were extraordinary examples of political bargaining. Payton, bearing a visible, though controlled rage over the betrayal of his appointed committee, was bound by his earlier statements to abide by their findings. What he needed most now was to save his face, not to mention his ass, from the Board of Trustees—McGrath in particular, who was not happy about having his name bandied about in print, and in court. In order to do this, he believed, he would have to accommodate me just enough so that I would accept reinstatement, and thereby agree to withdraw the lawsuit.

This was a terribly humiliating and painful process for Payton, who by this time had grown to dislike me even more than he disliked my politics. (If nothing else bothered him, he was offended by my appearance, which at the time was shabbier than the appearance of ninety-five per cent of the students. Payton was every inch the former diplomat, down to his highly polished, manicured fingernails, which he had a habit of checking out every thirty seconds or so. Why, I'm not sure. Perhaps he feared their sudden disappearance.)

I had mixed emotions about returning under any circumstances. My initial reaction to my firing had been relief, and originally I'd told myself I would not return. I had not really expected to be confronted with such a decision, and I agreed with the opinion of Professor Stanley Jarolem, the chairman of the investigating committee, who had stated in his footnote to the committee's report that my reinstatement would "not be in the best interest of the C. W. Post Center, nor in the best interest of Mr. Post." (I doubt that his statement was motivated by any great concern for my wellbeing.)

Finally, after four days of meetings, at which voices and tempers flared more than once, we emerged with a six-page document detailing an agreement acceptable to all, including the "withdrawal and termination of all legal action."

Agreeable to all, that is, but the Board of Trustees of Long Island University. That board included, besides Chairman McGrath, the William Zeckendorfs, Junior and Senior. McGrath, I had been told, headed a financial empire accumulated through Long Island banks and real estate. The Zeckendorfs, whose fortunes had been made in New York City real estate, including a well-known transaction involving properties in the United Nations area of Manhattan, had several years earlier declared bankruptcy, when their financial empire collapsed. The rest of the Board was equally far removed from academia.

The Board had always exercised tight control over the university. In fact, the trustees held WCWP's FCC license, and they were not about to relinquish control of the station to someone who had instituted legal action against them, thereby threatening to tarnish their already slightly bruised reputations. (This is simply an educated theory, though one which was shared by many on campus.)

Payton was the trustees' boy. One could almost see the strings attached to his limbs. But he had bungled this one, and the trustees' only recourse was to intervene themselves, which they would rather not have done, preferring instead to remain off stage, tugging away at Payton's strings.

I cannot recall exactly when I first learned of the Board's intervention, though I believe it was Payton himself who phoned, and in a voice barely able to conceal his joy, informed me that he would not, after all, be immediately able to implement our agreements and return the station to the air (it had now been silent for more than two weeks), until after the executive committee of the Board had met "to consider" the situation.

Again, the *Long Island Press* was on top of the story, thanks largely to the investigative work of *Press* reporter Richard Schiff, who had covered the developments all along, and who filed this report on December 16:

*The pressure [to hold off implementing the report], it was learned, was coming from the Long Island University Board of Trustees, who are involved in legal action in Brooklyn Federal Court dealing with the shutdown of the station and the firing of Post on Nov. 30.*

*The sixteen-man board is scheduled to meet today and is expected to take up the report of the twelve-member committee completed a week ago. It is not known whether an immediate decision will be made to abide by the report, but a party close to the situation said he would not be surprised if the board put off action until a Friday court appearance.*

*The court action . . . comes the same day the college closes for Christmas vacation. A delay in a decision by the Board of Trustees, and the court action, along with the holiday, could, therefore, work jointly to effectively keep the station off the air until New Year's.*

*. . . Though the Board of Trustees requested Payton late last week not to implement the report, its influence may have been felt*

even during the committee's investigation of the closing of the station [emphasis mine].

A student member of the Board, Francis L. McGrath, is the son of Long Island University's legal counsel, and the nephew of John P. McGrath, Board Chairman. The young McGrath was one of the three members of the student-faculty panel to sign a minority statement which read "We . . . do not feel that the best interests of the University community are served by the reinstatement of Mr. Post."

Post said he "honestly did not know whether McGrath might have served as a pipeline to the L.I.U. Board of Trustees," but added, "it's a possibility."

The Board, feeling confident of victory in court, apparently had decided to ignore the agreement to "terminate and withdraw all legal action." By going this route they could not only head off my reinstatement, but by drawing it out over the holidays could diffuse campus reaction and, they hoped, end, without the use of coercion or the threat of police action, the occupation of the Benjamin Abrams Communications Center. Thus, they reasoned, they could complete the job quietly, and, preferably, without any further press attention.

When Kunstler, who is certainly no stranger to the use of such tactics, heard of the Board's intervention, he seemed astounded. He proceeded to prepare supplemental affidavits, which included a summation of the student-faculty investigating committee's findings and recommendations, as well as the agreements for implementation reached by Payton, the temporary broadcasting panel, and myself.

The university's defense rested simply on the contention that the court had no jurisdiction in the case, that it was a matter to be brought before the FCC.

And so, on December 18, without permitting arguments on the merits of the case, The Honorable John R. Bartels upheld the defense's contention that the court lacked jurisdiction in the dispute, and denied our application for the immediate reopening of WCWP.

Kunstler immediately appealed the decision to the U.S. Court of Appeals, simultaneously filing suit once again in U.S. District Court, this time seeking "permanent injunctive relief." But neither of the cases would be heard until early in January. This news diminished the now slightly sagging spirits of what was left of the collective. The sit-in would have to continue through the two-week Christmas holiday. And so a dedicated core group of students and I continued the occupation of the building on the now nearly deserted campus, even as the college's maintenance people turned the heat down to the minimum necessary to preserve the building's foundations, if not ours.

On January 4, 1971, thirty-five days into the occupation of the Benjamin Abrams Communications Center and the day of the scheduled hearing on appeal, *Newsday* carried the following account of the holiday vigil:

*. . . the environs of the C. W. Post radio station might . . . be described as drab. But for the past month—ever since campus officials shut the station and fired its 26-year-old director—more than twenty student staffers and their supporters have been sleeping, eating and just plain living here.*

*Rather than becoming disheartened with their 34-day vigil—a show of resolve that deprived many of the youths of the delights of a Christmas vacation—the students appear more adamant than ever about reopening only under Steve Post . . . and operating under his new type of programing. College President Robert L. Payton . . . said last night, "Lord knows how long this whole thing will go on."*

*Inside the bare offices of the station, where sleeping bags cover the floors, student staffers continue to work on the briefs for their attorney . . . retyping, stenciling, mimeographing. When there is no work, there is time for television or snacks of hamburgers and ice cream brought in by friends and supporters. . . .*

*Last week . . . college officials offered to reopen the station for four hours a day [instead of the previous eight] on a programing schedule limited to music and news. "What kind of crap is that?" said Mark Indig, who directed a Sunday show on the station. . . .*

*"We're trying to create and experiment. . . . We'll stay here until the programing is opened up again."*

On that same day, January 4, a three-judge U.S. Court of Appeals panel refused to override Judge Bartels' order. Another hearing was scheduled for the following Monday, January 11, on the suit seeking a permanent injunction. The hearing, again, was to be held before Judge Bartels in U.S. District Court.

Despite the low morale of the collective, the occupation continued. On Thursday, January 6, I finally gave in to a painful abscess which had been developing near the base of my spine for more than a month and entered the hospital for surgery. Barring complications, the doctors said, I could be released on Monday, in time for the hearing.

On Friday, January 7, following surgery, I was greeted by my doctor with the disheartening news that I would require further surgery. But first I needed a week in the hospital to allow the first incision to heal. Surgery was tentatively set for Monday, January 18, a week after the scheduled second hearing in U.S. District Court.

At that hearing, Judge Bartels further heightened the suspense and prolonged the agony by reserving decision until Friday, January 15, or, he said, no later than Monday, January 18. (I was beginning to wonder whether the good judge and McGrath were not also pals with my surgeon.)

But on the afternoon of Tuesday, January 12, events took an extraordinary turn: President Robert Payton, on the direct orders of Board Chairman McGrath, ordered campus Security Chief George Sutton to evict the remaining protestors from the building. Sutton, after informing the students of this, set the deadline at three P.M., Wednesday, January 13. When asked by *Newsday* what actions the university might take if the students refused to leave he said, "a court injunction, college disciplinary procedures, and possible arrest are among the alternatives."

The weary students, now into the seventh week of what was unofficially the longest occupation in the history of United States campus activism, were genuinely caught off guard. Because the administration refused to be specific about what action they might take, it was difficult for the students to agree upon a course of action. They met and issued a statement saying only, "We will remain in the radio station until such time that our presence will serve no productive ends." When asked by *Newsday* what that meant, Mark Indig, acting as spokesman for the group said, "We certainly won't leave by three P.M. What we'll do after that is hard to say because we don't know what the administration is going to do." Another, unidentified spokesman for the group, summed it up well to the *Long Island Press:* "The administration, in deliberately trying to provoke a violent confrontation, has once again exhibited its scorn to those who believe in academic freedom."

It all seemed extraordinary, in view of the fact that the decision was still pending in the courts. "The net result," said William Kunstler in an interview with Julius Lester over WBAI, "will be young people who tried to use the system, who utilized the system's courts, and who are awaiting a decision now, as is the school, will find that the system retaliates by going into confrontations, and I just hope before that happens someone comes to their senses. . . . I'm hoping that someone realizes that all they're going to bring about is possible bloodshed and another bitter taste in the mouths of some American kids. . . . In fact," Kunstler continued, "the president [of the college] called the radio-station people who were sitting-in into his office the other day and told them in essence they were going to be punished for resorting to the courts, that instead of keeping it within the family of C. W. Post, they were now going to be punished because they had gone to the Federal Courts. . . ."

Frustrated and furious, I phoned the students from the hospital to see what was going on. Rumor had it, I was told, that if they refused to leave by the three-P.M. deadline the following day the college would seek a court injunction to have them forcibly evicted and possibly arrested. Though as a group they had come to no final decision, several of them were leaning toward resisting arrest. With my doctor agreeing to sign the appropriate release forms, I decided that if it came down to it I would return to the Benjamin Abrams Communications Center and join in the action.

At three P.M. Wednesday, January 13, as promised, Campus Security Chief Sutton arrived at the building. The confrontation proceeded as expected, with Sutton ordering them out, and the students refusing to leave. WBAI reporter Neal Conan, who was assigned to the story, asked Sutton what the next step would be. He replied, "We have several avenues open to us . . . the injunctive procedure, we also have a possibility of taking college disciplinary action which then could lead to criminal charges of trespassing. . . . So, I'm going to take [the students'] answer and return and report it to appropriate authority." The appropriate authority, according to Sutton, was Payton. And according to Conan's report, "*Mr. Payton was out of town today and I was told by the school's Public Relations Department that no decision was likely to be made until his return tomorrow.*"

Late in the afternoon I placed a call to Francis Koster, the campus ombudsman ("eyes and ears of the president"), who worked out of the president's office, who confirmed Conan's report. Though Koster couldn't say for sure exactly what the administration's next move would be, he did say it was likely to be the injunctive proceeding. In any event, *he could "assure me" that no further action would be taken until, "at the earliest," the following day, Thursday, January 14, when Payton was back on campus. Another call to the occupying students confirmed that they had been given precisely the same assurance from Koster and Dean Beeman, the Vice-President for Student Affairs.*

Thus assured that there would be no further attempts to remove the students that day, I felt my presence on the campus was not necessary. And, even if they filed an injunction the following morning, it could not be acted on until at least late that afternoon, or, more likely, Friday morning. In any event, I would have adequate time to check out of the hospital and make the less-than-one-hour journey to the campus. I wasn't certain, anyway, that my presence in the building would deter the administration from calling in police, but I knew it would be a psychological advantage to the students, and would likely attract media attention.

In the past seven weeks I had come to understand firsthand more about the reasons for student unrest and frustration, more about the mockery power makes of justice, than I could have in a lifetime of hearing about it and reading about it. (I had already taken a good deal of abuse about a satirical piece entitled "The Revolution Game," published in the *Realist* a year earlier. The piece was a satire of student demonstrators involved in an imaginary plot to occupy the "administration lounge of Bronx Community College.")

*At about seven P.M. that night, Security Chief Sutton, and the very same Dean Beeman and Francis Koster who two hours earlier had assured the students and me that no further action would be taken until the following morning "at the earliest," arrived at the station. Sutton made the following announcement: "I have orders from President Payton that unless you vacate the premises within fifteen minutes, everyone in the building will be under suspension."*

The students, again caught off guard, again misled by an administration and Board of Trustees who seemed willing to apply any tactic to maintain control, no matter what justice, campus or public opinion, or even the Federal Courts dictated, were stunned, demoralized, and completely exhausted. They filed bewilderedly out of the Benjamin Abrams Communications Center, for the first time in nearly two months, leaving behind sleeping bags, books, clothes, what was left of the pizza they had ordered for dinner, and, as it turned out, their hopes of creating and broadcasting over a free and open college radio station.

As the beleaguered students in the collective left the building they were given one last promise by Dean Alan Beeman: they would be provided with space somewhere on campus from which to continue their legal battle. This promise, like all the others, was never fulfilled. (Earlier we had been denied access to all office equipment and supplies, making it impossible to carry on the legal fight. But a supportive faculty member was willing to put her keys where her mouth was, and we slipped in and out of the humanities building at three and four in the morning, past campus security guards, and availed ourselves of the necessary items.)

I did not even learn about the action until the following morning. Mark Indig, sadly summing up for the students, was quoted in the January 14 edition of *Newsday*: "Most of the people in the collective are radio people, not political people; for most of us this is our first encounter with administrative pressure and repression, and a lot of us had to think twice about how far we would go." But, he added hopefully, "We're still going to fight to get our radio station back the way it was."

On Monday, January 18, Judge Bartels handed down his decision denying our pleas for "permanent injunctive relief," and fully supporting John McGrath. In that decision he stated that "the plaintiffs have failed to state a claim which requires vindication under the First or Fourteenth Amendments. . . . To hold that the licensing by the Federal Government of a broadcast frequency endows the recipient with the quality of an agency of the Federal Government would be an unreasonable interpretation of governmental action." To that, Kunstler replied in *Newsday*, January 21, "Ridiculous . . . a violation of all the principles of law. It's a terribly unfortunate thing when young people who rely on the law have this happen to them. It's ridiculous to say that the court does not have jurisdiction when there are twenty cases like it all over the place."

On Tuesday, January 19, a day later than planned (because, against doctors' orders and out of frustration, I had wolfed down a piece of seven-layer cake the night before), my rear was slashed for two hours, to add surgical insult to judicial injury, confining what was left of my body to the hospital for nearly a month.

On February 15 the administration returned WCWP to the airwaves, after a two-and-a-half-month silence, under the direction of William Mozer, formerly the station's chief engineer. It turned out that Mozer, who all along had expressed his support of the station's new direction and his indignation over the administration's action, had actually been negotiating for the position of director. Now Mozer would keep them happy. First, he denied all those students who had taken part in the occupation the right to return, in any capacity, to WCWP. Then he instituted a new format: four hours a day of music, with UPI rip-and-read newscasts on the hour. In five months, WCWP had come full circle.

My own immediate future rested on my stomach, in my home, to which I was confined for nearly two months following my release from the hospital, barely able to hobble from room to room with the aid of a cane, which I would have much preferred to wrap around the necks of Messrs. Payton and McGrath. Instead, I addressed a letter to Payton, which was printed in the school paper, *The Post Pioneer*. It said, in part:

*Congratulations! You have won the physical plant of a radio station. You have won the opportunity to turn that station into the voice of the Public Relations Department, or the Board of Trustees of Long Island University, or perhaps the voice of Robert L. Payton, or at best a bland imitation of a thousand other radio stations around the country. Good luck in your new endeavor.*

*In addition, you have provided a unique learning experience for me, for a number of students of all political beliefs, and for that segment of the public which has followed the WCWP affair. You have shown us, first hand, why it has become almost impossible to "speak truth to power" and have any hope of being heard, or of justice being done. You have shown us why many in our society have turned to violence in their frustrated efforts to seek just and humane change.*

*[I am addressing this letter to you] because at every turn you have had the opportunity to exercise moral leadership—to take a stand. At every turn you have chosen not to. Instead you have chosen the easiest, most self-serving course; you have chosen what you thought to be best for the continuing political career of Robert L. Payton. . . . And while you have paved the way for your own political future, you have helped destroy the faith of many in the integrity of that very political system of which you wish to become a part.*

*. . . Shortly before the student occupation ended, you told the students that the Board of Trustees would not implement the recommendations of the student/faculty investigating committee because they were being "punished" for taking this matter to the courts. Is this court case not an attempt to work through that very system in which you have led us to believe you have such great faith? What could possibly have been your motive this time for supporting the action of the board?*

Since the shutdown of the station and the beginning of the sit-in on Nov. 30, not a single illegal act has been committed. . . . No classes have been stopped, access to the building denied to no one. There has been no damage to the physical facilities of the radio station. The students' demands were reasonable ones: simply the implementation of the recommendations of a . . . committee appointed by you. Does this not fit your definition of "working within the system . . ."?

. . . It is unfortunate to see the administration of an academic community subordinate the interests of that community to self-serving motives. My brief association with the students at C. W. Post convinces me that they have maintained a high level of personal integrity despite the example set by your administration.

Mark Indig, who had been one of the creators and producers of *Satyricon, or A Tree Grows in Brookville,* the program which allegedly started it all some three months earlier, summed it up far less pompously, if no less indignantly, than I had in a letter published beside mine:

The Collective of WCWP wishes to award President Robert L. "Capitol Stuff" Payton their first annual "John McGrath Corruption Award" for extraordinary deceit and treachery in order to achieve the self-serving goals of power and prestige. Mr. Payton has also been cited for cowardice and repression above and beyond the call of liberal hypocrisy.

There are many things we admire about Bob, such as his famous springtime John Lindsay imitation, sports-jacket thrown over pinstriped shoulder, walking around campus shaking hands (pity there are no subways in Brookville), and his low-key Mario Procaccino approach to college administration, espousing a liberal philosophy, while ass-licking the Board of Trustees. Most of all we applaud what seems to be Bob's definition of Free Speech—whatever you can say to keep your job.

. . . We anxiously await [Bob's] rapid rise in the political arena, representing such underprivileged minority groups as misunderstood bankers and out-of-work ambassadors. If not, he can always have Alex Bennett's spot on WMCA. Sincerely,

Mark Indig and the WCWP Collective

Perhaps it was best summed up in brief letter addressed to Payton, a copy of which was sent to me. It was dated December 1, 1970, only one day after the station was silenced. It captures the meaning of the entire experience, at least for myself and the students. It is from Jerome W. Whitney of Shelton, Connecticut, whom I do not know. I assume he listened to WCWP:

Science is the search for reality—reality is truth.

What better scientific method could be found to teach your students a lesson about the reality and myth of due process in America than to summarily silence the student radio station and purge the entire staff.

Many of your students spend long hours taking notes in political science classes about the theory of the democratic process of America. Your coercive use of power has been an effective demonstration of how institutions in a "democratic" society operate in actual practice.

Thus you have taken your students out of the isolation of lecture hall abstraction into the classroom of daily life—the ordeal of reality—the experience of truth.

---

## NOTES FROM THE AUTHOR'S DIARY

FEBRUARY 21, 1971
RHINEBECK, NEW YORK

The euphoria is over now. The case is lost, the spotlight is gone, and my ass is killing me! The doctor says it will be another month before I can return to work. Not that I know what the fuck I'll do when I get back there. Goodman visited in the hospital and asked if I wanted to be program director. Said yes, but wasn't very lucid about how I'd handle the job. It's hard to believe, but I think he's leaning toward hiring that little woman. Josephson'll shit green. The fucking place will be overrun with cackling women ranting about their orgasms. . . .

Yesterday I sent my "farewell" letter to the college newspaper. All bullshit. There is really no simple way to sum up the meaning of this experience. Long ago I seem to have lost sight of where the issues became subordinate to my martyr's ego. There is no denying my genuine delight at having been the center of this controversy after so many years of standing in the shadows of WBAI's. I wonder how people like Kunstler contain those feelings? Maybe they don't. . . .

Now if I could only get up off my ass (what's left of it) and set it all down on paper before the feelings become distant, the events hazy. Before I lose contact with the "kids," as I know I will. Just a month ago I felt as close to them as I've felt to anyone, but already I feel myself pushing them out of my life. . . .

We were all too busy to keep diaries, so it will become impossible to reconstruct it all, at least in terms of the human relationships. Who really gives a shit anyway? . . .

————————

AUGUST 25, 1972
RHINEBECK

The exile seems to be doing me good. It's hard to believe, but I don't miss doing the program after all these years. Those last months were so fucking tough that I must have blocked the whole seven years just to keep from remembering the last few programs.

So I'm out of work now. First time in my life. No audience, no paycheck, no fan mail, no nothing. At least I'm working steadily on the book. It keeps my mind occupied, but it's scary. Now I really am writing about a bygone era.

C. W. Post chapter went smoothly. Thank god I saved all those papers and files. Real feeling of déjà vu, lock up here for three weeks recounting in detail and sequence the events of those few months. It was cathartic; I could almost smell the place at times. . . .

Spent a few days with Frank and Robin as a "reward" for finishing the Post chapter. Poor Frank, he's just moping around waiting for some busy-work to come along. We sit stoned for hours and commiserate, talking about the "good old days," like a couple of old vaudevilleans. Robin's into her "Good-Time Fascist" thing—country music and beer and fuck it all. . . .

What a pathetic crew we are!

## 12 "At This Time
## WBAI Concludes Another
## Day of Broadcasting"

It seems as though even before a dozen words of this work were committed to a writing surface, I was committed to a major anxiety attack over how it was to be concluded. But no satisfactory answer ever occurred.

In June 1972, tired, discouraged over the stagnation of my program, confused and pained by developments in my personal life, and somewhat disillusioned at the station's current political incarnation, I left *The Outside*, and WBAI, after nearly eight years. I announced on the air that I would be taking an "indefinite leave of absence," for that was the best way I knew to say good-by without having to say it. Like the last chapter of this book, the inevitable last program had, for years, loomed large and fearsome on future's list. To avoid the self-indulgent sentimentality inevitable in final farewells, especially one so privately public as this, I called it a "leave."

Then a peculiar thing happened. Two months later, at the end of August, Larry Josephson announced his departure from *In the Beginning . . .*, and the station, after nearly seven years. Several weeks later, near the end of September, Bob Fass relinquished *Radio Unnameable*, after almost a decade. It would not be long before Kathy Dobkin and Frank Coffee, two less-conspicuous "old-timers," would also pack up their transistors.

It seemed that WBAI, the child we had (so we believed) lovingly raised, was beginning to grow up and slip from our grasp.

Too much investment left us feeling cast out by those who had come along, like spoiled children, and taken our sacrifices for granted. Or was it we, like our parents, who expected their unquestioning devotion to us and our values?

By October 1972 I had been away and alone for nearly four months and still had given no indication to my audience, the station, or myself of whether or not I would return. I set down for my listeners some words of explanation. It was not intended for inclusion in this book, but as a short piece for WBAI's *Folio* which appeared under the title "Now I Know Why Lew Hill Committed Suicide." The thoughts in it seemed a bit too intimate for the random reader who, on whim or on sale, might grasp it from the mildewed out-of-print shelf of some obscure bookstore.

Though in January 1973 I returned to WBAI on a very limited basis, the essential conflicts remain unsolved. As the final draft of this manuscript prepares to pass into the anonymous hands that transform it from a mound of words into a bound book, I can find no more appropriate way to resolve it than by reprinting this piece, which seems to question all that has come before it.

---

*"Shit or get off the pot," my father used to say. I fully expect our institutional father, Ed Goodman, to repeat these words to me any day now. Each time we speak I set a new, later "tentative" date for my return to WBAI. Our conversations are punctuated with question marks, as is most of my life, and that of the station these days.*

*"What's going on at that place?" the more perceptive, or more paranoid of our regular listeners have been asking at every available moment, on every live phone-in program. No matter what the topic of the program, there are at least a handful of callers wanting to know the "real story" behind what must appear to be sweeping programing changes at WBAI.*

*It must seem all but impossible to the faithful—Josephson's gone after six years of "having puked my guts and soul out . . ."; Post and The Outside, after almost seven years, were* suddenly transformed into a summer squash and a dead rat one midnight in June; Fass, the father of us all and of free-form radio, is leaving Radio Unnameable after nearly a decade of innovation. Even Paul Gorman, the "new kid," is tossing in his shoes.

And the faithful have good reason to ask "why?" They (you) have supported this experiment in mass insanity over the years. Perhaps seventy-five per cent of our current subscribers signed up as a result of a plea, or simply the program of one of the "old gang."

Surely, the faithful imply, there must be a major internal political conflict, in the Pacifica tradition. It may, to a degree, be so—but it is not of the usual Pacifica variety. No single hot issue, not a mass purge, but simply the end result of an evolutionary political process—a carbon in microcosm of the larger political scene. And it is a situation I observe with particular personal despair, since I have supported, to a large extent, the current direction and management of the station, while ignoring, even ridiculing, the boy-who-cried-wolf–like warnings of my listener-sponsored contemporaries. Some will say my positions have been politically and personally expedient. I offer no defense, for I have never claimed to have much of a personal political philosophy—I am influenced by those around me whom I admire and respect at any given time, and I have supported many decisions which in retrospect seem like the wrong ones. But hindsight is easy, and I don't believe that the decision makers have been evilly motivated, or power hungry. We have all become captives of the rhetoric of this deplorably fragmented culture, though as an institution of innovation we should, perhaps, have exercised greater restraint.

But I did not intend here to write about the politics of WBAI, nor do I believe it is the primary reason for the seemingly sudden dissolution of WBAI's "old gang." It is, for me, and I believe for the others as well, a far more personal, complex anguish. Josephson, who has always had the envious gift of eloquence, summed up one aspect of it well: "I feel that I've said it all—many times over—and now need distance, quiet, and most of all, time— time to reflect, to replenish, to reorder, to grow." And if Josephson, who started with a mouthful, has said it all so many times in six years, I must have contracted lock-jaw after nearly seven.

But there must be some rational reason, a thread, that ties

On the air with *The Outside* at Pacifica's KPFT, Houston. ( PHOTO: PETER ZANGER )

James Irsay and friend in a penny arcade in Carolina Beach, North Carolina—placed here because these pictures are too beautiful to resist.

together this mutual feeling, this sudden sensation of being "burned out." Is the subculture stagnation of the seventies to blame? The death of input? Perhaps, to a degree in the cases of the others. But for myself it would be hypocrisy to place the blame externally. As my listeners well know, I have never taken much part in the culture/ subculture, much less have I relied upon it for input into the program.

The Outside has, more often than not, been the reflection of one man's internal conflict, expressed to an extent through the vehicle of humor, but identifiable still as human turmoil. At least that was the intent, if not always the result. Perhaps, as some have claimed, it was a selfish use of the public's airwaves. Pacifica's founders, after all, condemned the cult of the personality, and it took years to break down the internal resistance to this kind of programing, a resistance which, like a recurrent rash, seems to be erupting among WBAI's current epidermal layer. And I, perhaps even more than the others, have used my own personality as the prime source of energy from which the program flowed. My personal inadequacies became professional assets, though personal inadequacies they have most definitely remained.

Around the beginning of 1972 I found I was having greater and greater difficulty translating my own life into amusing and creative radio. Some (whose self-appointed mission, I believe, was to humor me) have said that I had outgrown the audience and the format. More likely though, recent intense turmoil and change in my own life have rendered me less capable of reflection through humor's always honest but often ugly mirror.

I strongly suspect that profound change in the personal lives of the others, though change different from my own, plays a role larger than they are willing to admit in their own sudden inability, or unwillingness, to translate life experience into creative radio. Like an athlete whose legs are giving out, we refuse to admit it to ourselves, we grasp at external straws for an explanation. And at WBAI what more available straw than politics?

It has been several months since I left the air on a regular basis. I am by myself, in my quaint little mobile home in the woods a hundred miles from New York, a hundred light years from the microphone. Since The Outside began, in November 1965, I have never been this long without an audience. Indeed, in my entire life, I have never been this long without human contact. Sometimes days pass without my even uttering a single word aloud. It is a strange, new kind of existence for one whose self-worth has always been measured by the opinions and reactions of others. I am my only audience for my witty remarks, the only victim of my put-downs, my only source of approval, the only target for my vindictive tantrums and petulance. I am learning to live with me, which I can assure you is no simple task.

The idea was to get away from WBAI for a while, to get away from the only world I have ever made or known. But there is nothing but WBAI for those of us whose lives have been touched and transformed by it. Dale Minor told me years ago to get out before it was too late. Only now do I understand what he meant. I am a "Pacifica Orphan"—beyond the station I have no identity. No geographical separation will change that. On December 1 I will celebrate (?) my eighth year with WBAI. Not simply working there, but living, sleeping, breathing, eating, and fucking there. Even my lovers and friends, with few exceptions over these eight years, have been BAI people. I have lost contact with and interest in almost everyone outside of this small world. I've begun to doubt whether I can even communicate on the most basic levels with anyone outside of this institution (an apt word). Can I live now without the nourishment of Pacifica's breasts? Where and how, at what seems still a relatively young age, will I find self? What practical, transferable training have I gained in these eight years of beautiful, albeit useless experience?

I verge on tears each time I think about Josephson's departure. Josephson, Fass, and I came along almost together, from a historical perspective. At times our egos clashed, we competed, we loved and hated each other, depending upon the time of day. There were times even when we didn't speak. But always there was, at the very least, a symbolic and unspoken brotherhood. Some, with justification, called it a political alliance. But we came together when the station was dying. We created a new kind of radio, a new audience and a new life for WBAI. We witnessed the corruption and death of our innovations at the hands of commercial radio. Am I to interpret our mutual restlessness, and Josephson's departure, as a finger pointing to the inevitable exit of an era? There seems no way to make that interpretation, no place far enough removed, no sufficient detachment. One must either go, as Josephson has, with a semblance of grace and the remnants of dignity, or return and face head on the personal and institutional struggles, which are too often inseparable.

Some time ago, in a naïve attempt to "ease my way out," I began writing extensively. But what did I choose to write about? WBAI, of course. Now, near the end of this untitled manuscript I've begun to recognize the futility of trying to communicate the "BAI experience" in print. There is an inherent idiocy in any attempt to transpose from one medium to another. But such revelations are a bit late in coming; I have a contract to fulfill. There seems no escape.

Perhaps that is what Josephson realized just in time—he has escaped, to the other end of the continent, at least. The "geographical cure." Fass, we all believed, sentenced himself to life years ago, though he too has loosened the noose. As for myself, all that I am certain of now is that The Outside is over. It had a long and reasonably healthy life, despite my occasional attempts at self-destruction, and finally passed away—not without pain, but quietly.

For now, I suppose, I will continue to put off a decision. I will take full advantage of the coddling of Pacifica's elders, until I am forced to either defecate or vacate. The suspense is killing me.